DORLING KINDERSLEY
—HANDBOOKS—

INSECTS

SPIDERS AND OTHER TERRESTRIAL ARTHROPODS

DORLING KINDERSLEY
—HANDBOOKS—

INSECTS

SPIDERS AND OTHER TERRESTRIAL ARTHROPODS

GEORGE C. MCGAVIN

Photography by
STEVE GORTON

Editorial Consultant
WILLIAM FOSTER
(University of Cambridge, UK)

A DORLING KINDERSLEY BOOK

LONDON, NEW YORK, MUNICH,
MELBOURNE, and DELH

Series Editor Peter Frances
Series Art Editor Vanessa Hamilton
Production Controller Michelle Thomas
DTP Designer Robert Campbell
Picture Research Andy Sansom
Senior Managing Editor Jonathan Metcalf
Senior Managing Art Editor Bryn Walls

Produced for Dorling Kindersley by

Editor Ann Kay
Art Editor Sharon Rudd

studio cactus

13 SOUTHGATE STREET WINCHESTER HAMPSHIRE SO23 9DZ

First published in Great Britain in 2000
by Dorling Kindersley Limited
80 Strand, London WC2R ORL

A Penguin Company

Copyright © 2000
Dorling Kindersley Limited, London
Text copyright © 2000 George C. McGavin

A CIP catalogue record for this book is
available from the British Library

ISBN 0-7513-0772-6

Reproduced by Colourscan, Singapore

Printed and bound by
Kyodo Printing Co., Singapore

see our complete catalogue at
www.dk.com

CONTENTS

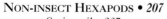

AUTHOR'S INTRODUCTION

Insects are the most numerous and successful creatures on Earth. They belong to a group of invertebrates known as arthropods, which are characterized by their jointed limbs, segmented bodies, and tough outer skeletons. Arthropods play an essential role in all of the world's major ecosystems. Although they are less conspicuous than other animals, if you look closely you will uncover their incredible variety and learn something about their extraordinary lives.

THE FIRST SIGNS of life on Earth were single-celled organisms that lived in the oceans around 3,500 million years ago. Jellyfish, simple worms, and other multicellular animals later evolved in the seas, followed by creatures with hard outsides, such as shellfish and trilobites – primitive arthropods. The early sea-dwelling arthropods were later to become the very first land-living animals, emerging from the oceans as scavengers about 420 million years ago – perhaps to escape aquatic predators. As land plants became more complex, they provided living space and resources for the increasing number of arthropod species – the most successful of all being the insects.

Today, invertebrate animals (those without a backbone) make up the majority of the world's known species; vertebrate animals (those with a backbone) account for less than three per cent. Within the invertebrates, the huge group Arthropoda eclipses all other groups, while arthropods are, in turn, dominated by the insects. It is estimated that there are about 10 quintillion – 10,000,000,000,000,000,000 – insects alive at any time.

UNWELCOME PESTS?

Arthropods are seen as troublesome pests by most urban-dwelling humans. Certainly, some are destructive. It is estimated that about 20 per cent of crops grown for human consumption are eaten by herbivorous insects. Insects also carry diseases that affect animals and human beings – approximately one in six people alive today is currently affected by an insect-borne disease. The venom of certain arthropods

△▷ INSECT ANCESTORS
Modern dragonflies evolved from species that begin to appear in fossils from around 250 million years ago. Very primitive dragonflies were flying through the lush, humid Carboniferous forests about 300 million years ago.

• *modern dragonfly is similar to primitive form*

150 million year-old fossil, formed in lithographic limestone

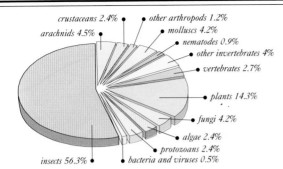

crustaceans 2.4%
arachnids 4.5%
other arthropods 1.2%
molluscs 4.2%
nematodes 0.9%
other invertebrates 4%
vertebrates 2.7%
plants 14.3%
fungi 4.2%
algae 2.4%
protozoans 2.4%
bacteria and viruses 0.5%
insects 56.3%

LIFE ON EARTH
Of all the species on Earth (represented left), 73.5 per cent are invertebrates, and most of these are arthropods. Insects – the most successful of all the arthropods in terms of survival and adapting to their environment – make up more than half of all species alive today. There are at least another four million insect species still to be named.

can be fatal, and many people have severe phobias about groups such as spiders and moths.

REAL BENEFITS
Most arthropods, however, are harmless. For example, less than one per cent of cockroach species – a much maligned insect group – are significant pests. Many people overlook the benefits that insects bring. Useful products derived from insects range from honey and silk to waxes, oils, natural medicines, and dyes. In many countries, insects

such as crickets, grasshoppers, grubs, and caterpillars still provide nutritious food for humans. Arthropods are also widely used in scientific research, helping us to understand genetics, physiology, and animal behaviour. Many insect species are vital plant-pollinators. Without them, many plants would die out and humans would lose a large proportion of their food supply.

PEST CONTROL
Predacious insects can help to control other, harmful, species. For example, ladybirds are efficient predators of soft-bodied species such as aphids.

• ladybird climbs up vegetation to find aphids

CHANGING THE COURSE OF HISTORY

Disease-carrying insects have made a major mark on history. Three world epidemics of plague, a flea-borne disease, killed millions and altered social structures as a result. Until the use of insecticides during World War II, twice as many people died of insect-borne diseases as from fighting. Most of Napoleon's army, which set out to conquer Russia in 1812, were killed by typhus spread by the human body louse. In the late 1800s, yellow fever carried by mosquitoes stopped the building of the Panama Canal for 15 years and killed 20,000 workers.

DEADLY MALARIA
It is estimated that malaria, a disease that is transmitted by certain mosquitoes, kills one human being every 12 seconds.

NATURAL BALANCE

The most fundamental role that the millions of arthropods play is in helping to maintain the balance of the Earth's ecosystems and food chains. These are complex networks that depend on energy from the Sun. The energy is "trapped" by green plants and converted to carbohydrate, which is then eaten by herbivores and converted into body tissue. The herbivores are then eaten by carnivores. Most food chains are dependent on insects as the majority of animals eat insects to survive and many would not exist without them. Birds, for example, are mostly insectivorous. A single swallow chick may consume around 200,000 bugs, flies, and beetles before it fledges, and even bird species that are seed-feeders as adults rear their young on a nutritious insect diet.

Animal droppings are food for certain beetles and flies, and many insects eat decomposing plant and animal matter. So insects are also helping to keep the Earth's supply of nutrients in circulation. Finally, although insects may cause serious damage to crops, they

long tongue used to catch insects

PREDATOR IN ACTION
Food chains are highly dependent on insects. Frogs, which are largely insectivores, are, in turn, hunted by larger predators.

can also control it. At least one-quarter of insect species are parasites or predators of other insects, and some are reared specially to control the numbers of agricultural pests.

A WORLD OF INSECTS

With around 1,500 families of terrestrial arthropods, it would be impossible to include them all in this book. We have chosen a broad range from around the world, including families because they are particularly important, common, or simply fascinating in some way.

▷ POLLINATORS
Without vital pollinators such as bees, many plants would not be able to produce fruit and seeds and so reproduce.

△▷ INSECT DEVASTATION
Since humans first started to cultivate crops, insect-borne plant diseases have caused the kind of damage shown right. One swarm of Desert Locusts (above) may contain up to 50 billion individuals, who could theoretically consume up to 100,000 tonnes of food a day.

HOW THIS BOOK WORKS

T HIS BOOK IS divided into 41 main sections, each covering a separate order of terrestrial arthropods. These sections are subdivided into entries that each describe the characteristics of a particular family with photographs of representative species. The family entries are arranged alphabetically by scientific name. Some of the order sections are divided into alphabetically arranged subdivisions, in which case this is explained in the introduction to the order. The sample page below shows a typical family entry.

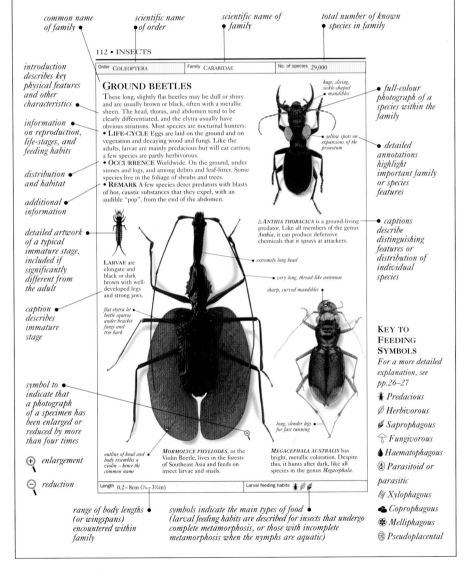

common name of family

scientific name of order

scientific name of family

total number of known species in family

112 • INSECTS

| Order COLEOPTERA | Family CARABIDAE | No. of species 29,000 |

introduction describes key physical features and other characteristics

information on reproduction, life-stages, and feeding habits

distribution and habitat

additional information

GROUND BEETLES

These long, slightly flat beetles may be dull or shiny and are usually brown or black, often with a metallic sheen. The head, thorax, and abdomen tend to be clearly differentiated, and the elytra usually have obvious striations. Most species are nocturnal hunters.
• LIFE-CYCLE Eggs are laid on the ground and on vegetation and decaying wood and fungi. Like the adults, larvae are mainly predacious but will eat carrion; a few species are partly herbivorous.
• OCCURRENCE Worldwide. On the ground, under stones and logs, and among debris and leaf-litter. Some species live in the foliage of shrubs and trees.
• REMARK A few species deter predators with blasts of hot, caustic substances that they expel, with an audible "pop", from the end of the abdomen.

huge, slicing, sickle-shaped mandibles

yellow spots on expansions of the pronotum

full-colour photograph of a species within the family

detailed annotations highlight important family or species features

△ ANTHIA THORACICA is a ground-living predator. Like all members of the genus *Anthia*, it can produce defensive chemicals that it sprays at attackers.

captions describe distinguishing features or distribution of individual species

detailed artwork of a typical immature stage, included if significantly different from the adult

LARVAE are elongate and black or dark brown with well-developed legs and strong jaws.

extremely long head

very long, thread-like antennae

sharp, curved mandibles

caption describes immature stage

flat elytra let beetle squeeze under bracket fungi and tree bark

symbol to indicate that a photograph of a specimen has been enlarged or reduced by more than four times

⊕ enlargement

⊖ reduction

outline of head and body resembles a violin – hence the common name

MORMOLYCE PHYLLODES, or the Violin Beetle, lives in the forests of Southeast Asia and feeds on insect larvae and snails.

long, slender legs for fast running

MEGACEPHALA AUSTRALIS has bright, metallic coloration. Despite this, it hunts after dark, like all species in the genus *Megacephala*.

| Length 0.2–8cm (¹⁄₁₆–3¼in) | Larval feeding habits ✹ ⊘ ⸙ |

range of body lengths (or wingspans) encountered within family

symbols indicate the main types of food (larval feeding habits are described for insects that undergo complete metamorphosis, or those with incomplete metamorphosis when the nymphs are aquatic)

KEY TO FEEDING SYMBOLS

For a more detailed explanation, see pp.26–27

✹ Predacious

⊘ Herbivorous

⸙ Saprophagous

⸸ Fungivorous

◖ Haematophagous

⊛ Parasitoid or parasitic

⸙ Xylophagous

● Coprophagous

⊛ Melliphagous

⊛ Pseudoplacental

WHAT IS AN ARTHROPOD?

LIVING THINGS ARE GROUPED by biologists into five major divisions known as kingdoms, the largest of which is the animal kingdom. The kingdoms are in turn divided into groups called phyla. Arthropods form the largest single phylum in the animal kingdom. They comprise an incredibly diverse group, ranging in size from mites a fraction of a millimetre long to the vast Japanese Island Crab,

Macrocheira kaempferi, which can grow up to 4m (12ft) across. Arthropods are found in every habitat on Earth, from the depths of the oceans to the highest peaks, from arid deserts to the most humid rainforests, as well as in highly populated urban areas. Insects are the only winged arthropods, and are the most successful in terms of survival.

Below is a simple "tree" showing the sub-divisions of the phylum Arthropoda.

Phylum
A major sub-division of a kingdom (in this case the animal kingdom, or the kingdom Animalia).

> **ARTHROPODA**
> *Animals with a tough outer skeleton, segmented bodies, and jointed legs*

Sub-phylum
A major sub-division of a phylum (in this case the phylum Arthropoda).

> **MANDIBULATA**
> *Arthropods with antennae, and highly modified jaws for biting or chewing*

Super-class
A sub-division of a sub-phylum, composed of classes of animals that share fundamental characteristics.

> **HEXAPODA**
> *Mainly terrestrial, 6-legged arthropods, with 2 antennae*

SPRINGTAIL
Isotoma viridis
(p.207)

Class
A group made up of orders of animals that share similar characteristics.

> **INSECTA**
> *The only winged arthropods*

> **NON-INSECT HEXAPODA**
> 3 classes:
> COLLEMBOLA
> PROTURA
> DIPLURA

STINK BUG
Eurydema dominulus
(p.92)

Order
A group made up of closely related animal families.

29 orders	3 orders

Family, genus, species
A family is made up of similar species. A genus is made up of closely related species.

949 families 1,000,000 species	31 families 7,700 species

MAIN CHARACTERISTICS OF ARTHROPODS

Arthropods share a number of common features:
• Bilaterally symmetrical bodies.
• A protective, rigid outer exoskeleton (or cuticle) made of a tough material called chitin. The muscles are attached to this exoskeleton, which is moulted from time to time as the animal develops through its life stages.
• Pairs of jointed legs, which arise from the body segments.
• Body segments that are arranged to form a few main sections, the most common being the head. Myriapods have a head and trunk; crustaceans and hexapods have a separate head, thorax, and abdomen. In arachnids, the head and thorax are fused to form a single segment known as the cephalothorax.

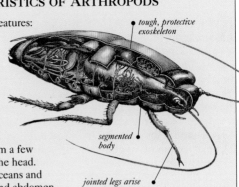

tough, protective exoskeleton

segmented body

jointed legs arise from body segments

PORCELLIONID
Porcellio scaber
(p.212)

CYLINDER MYRIAPOD
Julus species
(p.242)

CHELICERATA
Arthropods with pincer-like mouthparts and no antennae

CRUSTACEA
Mainly aquatic arthropods with gills and 4 antennae

MYRIAPODA
Arthropods with 9 or more pairs of legs and 2 antennae

FUNNEL WEAVER
Tegenaria gigantica
(p.228)

6 classes:
REMIPEDIA
CEPHALOCARIDA
BRANCHIOPODA
OSTRACODA
MAXILLOPODA
MALACOSTRACA

4 classes:
PAUROPODA
SYMPHYLA
CHILOPODA
DIPLOPODA

3 classes:
ARACHNIDA
PYCNOGONIDA
MEROSTOMATA

37 orders

16 orders

14 orders

540 families
40,000 species

144 families
13,700 species

470 families
76,500 species

WHAT IS AN INSECT?

MANY PEOPLE CONFUSE insects and other arthropods. Insects, like all arthropods, have jointed legs and a hard cuticle, but unlike others they have only six legs and, usually, wings. The word "insect" is derived from Latin, meaning "to cut into", and refers to the separate sections that make up an insect – the head, thorax, and abdomen. The head carries the mouthparts, antennae, and eyes. The thorax has three segments, with legs and sometimes wings. The abdomen has up to 11 visible segments and may carry terminal "tails" (cerci).

THE PARTS OF AN INSECT

Through the process of evolution, the basic insect body parts have become modified in a variety of ways in different insects. For example, the mouthparts may be adapted either to bite and chew or to suck liquids such as blood, nectar, or plant juices. The antennae are vital sensory organs that can respond to chemicals, such as a mate's sexual odours, or to physical stimuli, such as hosts moving deep within plant tissue. Insects' legs are modified for jumping, digging, swimming, catching prey, and even hearing and singing. Wings are not just for flying – they can be tough and protective, reflect the sun, or act as air stores. Their coloration may act as camouflage, or be used to signal to mates or scare off enemies.

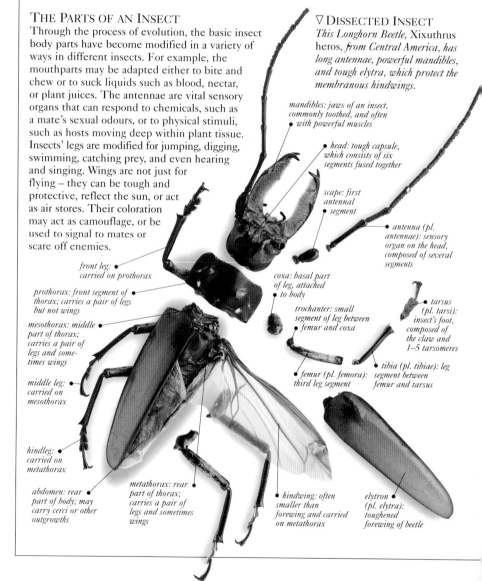

▽ DISSECTED INSECT
This Longhorn Beetle, Xixuthrus heros, from Central America, has long antennae, powerful mandibles, and tough elytra, which protect the membranous hindwings.

mandibles: jaws of an insect, commonly toothed, and often with powerful muscles

head: tough capsule, which consists of six segments fused together

scape: first antennal segment

antenna (pl. antennae): sensory organ on the head, composed of several segments

front leg: carried on prothorax

prothorax: front segment of thorax; carries a pair of legs but not wings

mesothorax: middle part of thorax; carries a pair of legs and sometimes wings

middle leg: carried on mesothorax

hindleg: carried on metathorax

abdomen: rear part of body; may carry cerci or other outgrowths

metathorax: rear part of thorax; carries a pair of legs and sometimes wings

coxa: basal part of leg, attached to body

trochanter: small segment of leg between femur and coxa

femur (pl. femora): third leg segment

tarsus (pl. tarsi): insect's foot, composed of the claw and 1–5 tarsomeres

tibia (pl. tibiae): leg segment between femur and tarsus

hindwing: often smaller than forewing and carried on metathorax

elytron (pl. elytra): toughened forewing of beetle

INSIDE AN INSECT

An insect's central nervous system consists of a brain connected to nerve masses called ganglia. The peripheral nervous system is composed of sensory nerves that gather information from sensory receptors and motor nerves that control the muscles.

The respiratory system consists of a network of tubes. Air is taken in through openings, called spiracles, on the abdomen or thorax. The immature stages of aquatic species take in air through gills.

The circulatory system is an open one, where the organs are bathed in a fluid called haemolymph that transports nutrients and waste around the body.

The digestive system is an open-ended tube with areas for grinding and storing food, producing enzymes, and absorbing nutrients.

To reproduce, males typically transfer sperm to the female's sperm store via a penis (aedeagus), and eggs are fertilized as they pass down the female's oviduct.

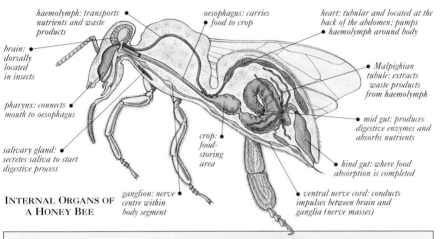

haemolymph: transports nutrients and waste products

oesophagus: carries food to crop

heart: tubular and located at the back of the abdomen; pumps haemolymph around body

brain: dorsally located in insects

Malpighian tubule: extracts waste products from haemolymph

pharynx: connects mouth to oesophagus

mid gut: produces digestive enzymes and absorbs nutrients

salivary gland: secretes saliva to start digestive process

crop: food-storing area

hind gut: where food absorption is completed

INTERNAL ORGANS OF A HONEY BEE

ganglion: nerve centre within body segment

ventral nerve cord: conducts impulses between brain and ganglia (nerve masses)

NON-INSECT HEXAPODS

Insects belong to a group of animals called hexapods (see pp.10–11). This group also includes three classes – collectively known as non-insect hexapods – that are generally regarded as being distinct from insects: the Diplura (diplurans), Protura (proturans), and Collembola (springtails).

Most non-insect hexapods live in soil or leaf-litter. None have wings, and some even lack eyes and antennae. The major difference between these hexapods and insects, however, concerns their mouthparts. Unlike insects, the non-insect hexapods have their mouthparts enclosed within a pouch, which is located on the underside of the head. When in use, the mouthparts are pushed out of the pouch to scrape, bite, or pierce the food.

SPRINGTAILS

These arthropods are the most abundant and widespread of the non-insect hexapod classes. They are either elongate or rounded – like the springtail shown here (See pp.207–11).

TYPES OF INSECT

Insects can be divided into three groups, depending on the way that they develop during their lifetime. Primitive, wingless insects, such as silverfish, develop to adulthood by moulting periodically throughout their life. Winged insects change either by a gradual process, called incomplete metamorphosis, or undergo a more sudden transformation, called complete metamorphosis, which involves a pupal stage (see pp.20–23).

The very first winged insects developed by gradual metamorphosis. A pupal stage did not evolve until the Permian period (290–245 million years ago), perhaps in response to climatic conditions (the pupa made it possible for insects to survive a period of cold). An increasing degree of tissue reorganization within the pupa also meant that larval stages were no longer just miniature versions of adults. Larvae became "eating machines" and adults "breeding machines". The success of the pupal stage can be seen clearly today. Eighty-five per cent of all living insect species develop in this way, and the majority of those that do belong to one of four large, successful orders: the Coleoptera, Diptera, Lepidoptera, and Hymenoptera.

△ MOST PRIMITIVE
The first insects were wingless scavengers that appeared more than 400 million years ago. The most primitive insects alive today, the bristletails (above) and silverfish, are similar in function and appearance.

△ MOST ADVANCED
Insects of the order Hymenoptera (see pp.178–206), such as bees (above), are considered to be the most advanced. Many live in colonies, often with castes that perform separate tasks.

BIGGEST
Insects in prehistoric times were much larger than they are today. Large species, however, still survive, and the spider-hunting wasp, shown here, which grows up to 7cm (2¾in), is among the biggest species alive today. The smallest species could sit on its foot.

SMALLEST
Some parasitic wasps (above) are among the tiniest insects on Earth, with a body length less than 1mm (½in).

WINGS AND FLIGHT

One of the key factors in the success of insects as terrestrial species was the evolution of flight. Insects were the first animals to take to the air, and this ability enabled them to evade enemies and find food and mates efficiently. Insects had evolved wings before the Carboniferous period (350–290 million years ago), but these early fliers were not able to fold their wings back along their bodies. By the middle of this period, some insects had evolved this ability, and it allowed them to use a far greater range of microhabitats, such as cracks and crevices in dead wood, inside leaf-litter, or under stones. It also meant that they could hide from predators. The descendants of these species were highly successful, and today it is only dragonflies and mayflies that cannot fold their wings in this way.

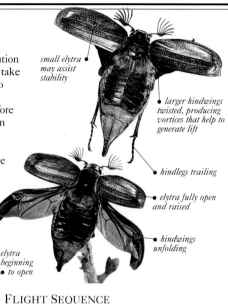

small elytra may assist stability

larger hindwings twisted, producing vortices that help to generate lift

hindlegs trailing

elytra fully open and raised

hindwings unfolding

elytra beginning to open

FLIGHT SEQUENCE

These steps show the cockchafer beetle preparing to fly and then in full flight. Unlike birds, insects need to warm up their flight muscles before they can take off. They do this by basking in the sun or vibrating their wings.

THE SECRETS OF SUCCESS

Throughout their evolution, several factors have combined to make insects the most successful of all species on this planet. Today, they make up over half of the species alive. There are several reasons for their success, mainly their ability to fly and reproduce quickly, their generally small size, and their protective cuticle (external exoskeleton) and their insulated central nervous system.

FACTOR	EFFECT
CUTICLE (EXTERNAL EXOSKELETON)	*Tough and waterproof, the cuticle helps to protect insects from predators and also from dehydration.*
FLIGHT	*This allows insects to escape from enemies, to find new habitats and food sources rapidly, and to establish new colonies.*
FAST REPRODUCTION	*Insects evolve at a high rate and adapt quickly to changing environmental conditions.*
INSULATED CENTRAL NERVOUS SYSTEM	*Insulation of the central nervous system allows nerves to work efficiently and also allows survival in hot or dry places.*
SIZE	*Small size allows utilization of a wide range of microhabitats – a tree, for example, may support hundreds of insect species.*

WHAT IS AN ARACHNID?

ARACHNIDS, which include spiders, scorpions, ticks, and mites, differ from insects essentially in that their bodies are divided into two rather than three segments. Their ancestors were marine, scorpion-like creatures, which flourished during the Silurian Period (435–400 million years ago); some of these were more than 1m (39in) long. The marine species died out about 250 million years ago, but their descendants have been highly successful on land.

THE PARTS OF AN ARACHNID

An arachnid's body is divided into two parts. The head and thorax are fused together, forming a cephalothorax, or prosoma, which is joined to the abdomen, or opisthosoma. In some, the abdomen is segmented and may have a tail-like extension; spiders' abdomens contain silk glands. An arachnid's cephalothorax has six pairs of appendages. The first pair (chelicerae) may be pincer- or fang-like, and are used mainly for feeding. The second pair (pedipalps) have several functions, including capturing prey and fertilizing the female, and may be leg-like or enlarged with terminal claws. The other four pairs are walking legs, although the first pair may also carry sensory organs. Gases are exchanged through the trachea or special respiratory organs called book lungs. Most arachnids digest their food outside the body using enzymes, which are pumped into or poured over food. The resulting liquid is then sucked up.

fourth walking leg

third walking leg

second walking leg

abdomen

cephalothorax

first walking leg

chelicera • pedipalp

MEXICAN RED-LEGGED TARANTULA

▷ INSIDE AN ARACHNID
The cephalothorax houses the brain and sensory organs, as well as the sucking stomach and venom gland. The abdomen is concerned with digestion, gaseous exchange, reproduction and – in spiders (as illustrated here), pseudoscorpions, and some mites – the production of silk.

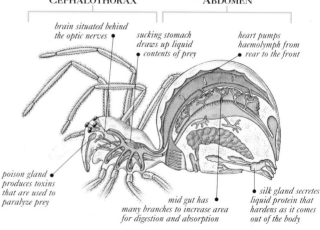

CEPHALOTHORAX   ABDOMEN

brain situated behind the optic nerves •

sucking stomach draws up liquid • contents of prey

heart pumps haemolymph from • rear to the front

poison gland produces toxins that are used to paralyze prey

mid gut has many branches to increase area for digestion and absorption

silk gland secretes liquid protein that hardens as it comes out of the body

◁ HARVESTMAN
The cephalothorax and abdomen in harvestmen are joined in such a way that they look as though they have only one body section.

▷ TICK
The protruding structure at the front of a tick's body houses barbed mouthparts used to penetrate the host's skin.

TYPES OF ARACHNID

Arachnids are a large and diverse group. They are divided into 11 orders, each of which has characteristic features. Sun-spiders, for example, have massive, forward-facing chelicerae. Scorpions are recognizable by their long abdominal "tails", bearing stings, and their large, claw-like pedipalps. Whip-scorpions also have large pedipalps, but they are not claw-like, and the long, whip-like tail is without a sting. Perhaps the biggest variation in appearance is seen in the spiders and the mites. Spiders vary from tiny money spiders with turreted, eye-bearing extensions on the cephalothorax to huge, hairy species, known as tarantulas. The huge number of species that make up the mites and ticks vary from gall-forming mites, which are probably the smallest arthropods in the world at less than 0.1mm (under ⅟₂₈in) long, to blood-feeding ticks, which can be more than 30mm (1¼in) long. Some have slender or flattened bodies that allow them to fit inside a human hair follicle or burrow through skin layers.

SPIDERS' WEBS

Spiders produce silk to wrap their eggs in, and for lining burrows and making shelters, but the most well-known use is for capturing prey. (Not all spiders catch prey using silk; some simply rely on good eyesight and stealth.) Web-making spiders have evolved various ingenious prey-capturing techniques, several of which are shown below.

ORB WEB
Spirals of sticky silk are constructed across open spaces. Some webs can be strong enough to catch birds.

TRAP DOOR
This is a silk-lined tunnel with a hinged lid, to provide shelter and protection as the spider waits for prey.

CAST WEB
Some spiders make small webs that they hold in their legs and throw over passing prey.

COB WEB
Tangles of silk seen in buildings may be made by daddy-long-legs. Other species make cob webs in vegetation.

young scorpions cluster together on mother's back for protection •

PARENTAL CARE

Many arachnids, including some harvestmen and ticks, show parental care by guarding their eggs from predators. Scorpions, whip-spiders, whip-scorpions, and some spiders carry their young around on their backs for a while after they emerge from the egg sac or brood chamber.

WHAT IS A CRUSTACEAN?

DIVERSE IN APPEARANCE, crustaceans range from water-fleas, barnacles, and sand-hoppers to shrimps, crabs, and lobsters. The group ranges in size from microscopic plankton to giant lobsters that reach lengths of more than 75cm (30in). They are primarily aquatic, and typically have a distinctive hardened carapace. They occur in freshwater and marine habitats throughout the world.

Some crustacean species have adapted to life on land; woodlice, for example, are exclusively terrestrial and are common and widespread. Most crustaceans are scavengers, but there are predatory and herbivorous species, and some, such as barnacles, filter minute particles of food from the water using modified, strainer-like legs.

THE PARTS OF A CRUSTACEAN

The carapace of crustaceans is similar to the exoskeleton of other arthropods, but is often strengthened with deposits of calcium carbonate. The head and thorax are often covered by a single carapace. Crustaceans have a second pair of antennae, and their appendages are specialized for a number of functions, ranging from collecting sensory information to movement, respiration, and egg brooding. Their appendages are double-branched, a basal portion bearing an inner part, which is used for walking, and an outer part, which is used for swimming.

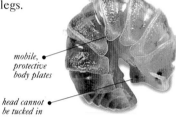

mobile, protective body plates

head cannot be tucked in

cuticle protects against dehydration

△ARMADILLIDIUM ALBUM
Woodlice roll into a ball when threatened but cannot tuck their heads in, unlike pill millipedes.

◁PILL WOODLOUSE
Descended from an aquatic species, pill woodlice still prefer damp places.

MARINE CRUSTACEANS

Most crustaceans are abundant in the sea and belong to the group Malacostraca. This includes the more familiar crab, shrimp, and lobster species. The front of the carapace often extends to form a projection, the eyes are stalked and compound, and the abdomen ends in a tail-like telson. In crabs, the abdomen is short and tightly curled to fit under the broad carapace.

△COMMON LOBSTER
In many larger species, such as the common lobster (above), the first pair of thoracic legs is enlarged, with strong claws. These are used for defence, handling food, and courtship.

swimming legs

thoracic limbs covered with fine hair

compound eye

KRILL

WHAT IS A MYRIAPOD?

MYRIAPODS ARE SIMILAR to insects in many ways, and the two groups are considered to be close relatives. Both have mandibles and lack the branched legs and second pair of antennae found in crustaceans. They also have some similar internal organs such as the tracheal system and malpighian tubes.

However, some evidence suggests that insects may be closer to crustaceans, and that the legs and antennae of insects have evolved differently in response to life on land.

THE PARTS OF A MYRIAPOD

These typically elongate, terrestrial animals are distinguished from all other arthropods by their numerous pairs of legs, and by having a trunk that is not divided into a separate thorax and abdomen. They have one pair each of antennae and mandibles. The cuticle of a myriapod is not as waterproof as an insect's, and myriapods cannot close off the spiracular openings to the tracheal system; as a result, they are mostly confined to humid microhabitats, such as soil and leaf litter, and are usually nocturnal. Differing gaits, determined by leg length and number, are seen in fast-running and burrowing species.

• *first segment behind head has pair of poison claws*

• *trunk segments fused in pairs*

• *two pairs of legs per segment*

△ **CENTIPEDE**
Typically fast-moving and predacious, centipedes have trunk segments that carry one pair of legs each.

◁ **MILLIPEDE**
These are typically slow-moving, burrowing species. Most of their trunk segments are fused in pairs, called diplosegments, each bearing two pairs of legs.

◁ **SYMPHYLAN**
Closely related to the centipede, this soft-bodied creature lives in soil and leaf-litter. It has fairly long antennae and 12 pairs of legs.

▷ **PAUROPOD**
A close relation of the millipede, this myriapod inhabits leaf-litter and soil. It has a soft body, short, branched antennae, no eyes, and nine pairs of legs.

SELF-DEFENCE

Centipedes use their poison claws for self-defence (the bite can cause vomiting and fever in humans). Pill millipedes can roll into a ball with their head tucked in under the last tergite (abdominal plate).

tergite extends to cover legs • *head tucked in*

PILL MILLIPEDES

LIFE-CYCLE

ALL ARTHROPODS must shed their exoskeleton at intervals in order to grow, but the development from egg to adulthood varies between the different groups. Myriapods and arachnids moult throughout their lives, and immature stages typically look like smaller versions of the adults. Insects, however, with the exception of bristletails and silverfish (see p.23), change their appearance from the immature to the adult stage. In more primitive insects, change is gradual and the metamorphosis is described as "incomplete" (see below); in advanced insects the change is often extremely dramatic and the metamorphosis is known as "complete" (see pp.22–23).

INCOMPLETE METAMORPHOSIS

In insects that undergo incomplete metamorphosis, the immature stages are called nymphs. The nymphs look very similar to adults but lack wings and reproductive structures. Wings develop gradually on the outside of the body, inside wing buds or pads. After a series of moults – the precise number varying between species – the final moult to adulthood occurs with the expansion of the wings. In aquatic orders, such as dragonflies and damselflies, the nymphs are less like the adults.

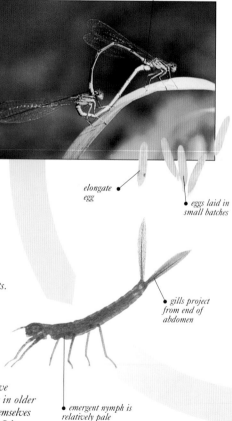

tip of female's abdomen joins male's genitalia

male holds female using clasping organs at the end of its abdomen

1. MATING

In the damselfly species shown here, Coenagrion puella, *the male transfers sperm from its primary genital organs on the ninth abdominal segment to secondary genitalia on the third abdominal segment. The male clasps the female behind the head, while the female bends her abdomen around to join with the secondary genitalia of the male. Sperm is transferred via the penis to the female's sperm storage organ. Eggs are laid inside aquatic plants.*

elongate egg

eggs laid in small batches

2. EMERGENT NYMPH

After the pale nymph emerges from the egg, it develops through a series of stages called instars – the number of which varies between species, and according to temperature and food supply. The first few instars do not have visible wing pads. Pads become more noticeable in older nymphs. Although predacious, the nymphs are themselves prey to many creatures, such as water beetles and fish.

gills project from end of abdomen

emergent nymph is relatively pale

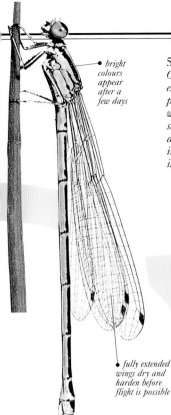

bright colours appear after a few days

fully extended wings dry and harden before flight is possible

5. ADULT DAMSELFLY
Once free of the nymphal skin, the adult can expand its abdomen and wings to full size by pumping haemolymph into them. The cuticle will harden in a few hours, but it will be several days before the bright adult coloration develops. The female is not ready to mate immediately; it feeds for a week or two while its ovaries mature.

4. EMERGING ADULT
Anchored by the nymphal skin, the emerging adult pushes upwards and takes hold of a stem using its legs. Pulling itself forward, it gently eases the wings out of the wing buds and then removes its abdomen. At this stage the thorax has not acquired its final shape, the body is soft, and the wings are crumpled and unexpanded.

abdomen is final body part to emerge

empty nymphal skin remains attached to stem

swollen thorax caused by pressure of haemolymph

nymph uses claws to crawl up plant stem

relatively dark coloration

3. FINAL INSTAR
When a nymph is fully grown, it begins to emerge from the water for increasingly long periods of time and eventually leaves for good. An increase in haemolymph pressure in the thorax causes a split along the back of the nymphal cuticle between the wing pads. The adult's head and thorax separate from the old skin first and emerge.

COMPLETE METAMORPHOSIS

In the insects that develop by complete metamorphosis, the immature stage, called the larva, looks completely different from the adult. The larvae of some flies are known as maggots; the larvae of many beetles are known as grubs; and the larvae of butterflies and moths are called caterpillars. The larvae feed continuously and go through a number of moults until the final larval stage is reached. They then stop feeding and search for a suitable place to pupate. In the pupal stage, the reorganization and transformation of the larval tissues into adult structures takes place. The tissues of the immature insect are broken down and small groups of cells called imaginal discs, which have been present since the egg first hatched, grow and develop into adult organ systems. To protect the pupa, the final larval stage often spins a cocoon or makes a cell out of soil particles or chewed wood fibres. The pupae of some species have moveable jaws and are able to defend themselves to a certain extent. The adult frees itself from the pupal skin and/or cocoon by using its jaws, legs, or by swelling parts of the body.

1. MATING

Courtship may involve the production of sexual odours, sounds, and even light displays. In the ladybird species shown here, Coccinella septempunctata, *the male clings to the back of his mate. Sperm may be transferred in a matter of minutes, but by maintaining hold for a longer time, the male makes sure that other males do not mate with the same female.*

male clings to •
back of female

• clusters of eggs
laid on foliage

2. EMERGENT LARVAE

The eggs are laid in relatively small batches on the leaves of plants, and after about one week the minute first instar larvae emerge. The cuticle is soft at first but soon hardens and darkens. The larvae must find suitable soft-bodied prey, which in this species are aphids of various kinds.

• larva pulls itself
free of egg case

empty shell remains •
stuck to leaf surface

spines and
projections
on body •

3. FINAL LARVAL STAGE

The dark-coloured, elongate larvae have well-developed spines and projections on their bodies, and strong legs. They can be found on stems and the undersides of leaves wherever there are aphids to eat. A single ladybird larva can eat many hundreds of aphids during its larval development. Its pale spots signal that it is distasteful to potential predators.

orange spots •
deter predators

*red coloration
darkens over
several days* •

6. ADULT LADYBIRD

*The distinctive, hemispherical adult is protected from
predators, such as birds, by its bright warning colours
and the ability to exude distasteful liquids from its leg
joints. Like the larvae, the adults feed on aphids and
other soft-bodied insects and are useful in biological
control of pests. Many ladybird species overwinter in
groups in sheltered places outside or in buildings, and
emerge to lay their eggs in the springtime.*

*elytra are soft and
vulnerable* •

5. NEW ADULT

*After a week or so, pupation is complete. The
pupal cuticle splits down the back, and the
pale, soft adult emerges. During the hour or two
after emergence, the elytra must be raised and
the hindwings expanded from underneath
and hardened before they can be folded
away again ready for flight. The bright
colours and contrasting spots of the adult
beetle will take a couple of days to appear.*

*black thorax
with pale yellow
spots in this
species* •

pale coloration •

4. PUPATION

*After about four weeks, depending on
conditions such as temperature and
the supply of food, the fully grown
larva pupates. It attaches itself
to the underside of a leaf and
sheds its last larval skin. The
larva remains immobile and
the pupal cuticle underneath
hardens and becomes dark.*

*cuticle hardens and
darkens* •

AMETABOLOUS INSECTS

In insect species where there is no change
in shape and where moulting continues
even after sexual maturity is reached,
development is said to be ametabolous.
Only two orders, bristletails and silverfish,
comprising less than 0.1 per cent of all
insect species, develop in this way. In
addition to thoracic legs, these primitive,
scavenging species have short appendages
on some of their abdominal segments.

SILVERFISH
*The streamlined shape gives
the silverfish its name.*

THE SENSORY SYSTEM

ALTHOUGH MOST ARTHROPODS are very small creatures, they possess surprisingly sophisticated sensory systems that allow them to respond appropriately to a wide range of internal and external stimuli. Arthropods are able to receive visual, chemical, and mechanical cues, many have temperature and humidity sensors, and some can detect magnetic fields and infra-red radiation.

HOW INSECTS SEE

In most insects, except for some cave-dwelling species, some parasitic groups, and the worker castes of some ants and termites, the main visual organs are compound eyes (see below). In day-flying species, the image received is made up of numerous tiny spots of differing light intensity. Night- and dusk-flying insects have eyes that are adapted to dim light conditions, although the images they form are not as sharp as in day-flying species. Many adult insects and some immature ones have simple eyes, called ocelli, either instead of or as well as the compound eyes. Simple eyes respond to light or dark only, and are important in determining certain behavioural rhythms, such as when to forage for food or hibernate. Colour vision occurs in all orders of insects. Generally, insects see better at the blue end of the spectrum rather than the red. Some insects are sensitive to ultra-violet light.

dark area guides bee to nectar

flower in visible light

△ A BEE'S VIEW

Flowers may have very distinctive patterns, called nectar guides, which reflect ultra-violet light and are visible only to bees and some other species. This photograph was taken with UV-sensitive film.

▽ EYES OF A HUNTER

Eyes are important for finding food and mates. Aerial predators often have larger eyes and sharper vision than other insects, sometimes with the eyes covering the entire surface of the head. Insects that mate in swarms also tend to have big eyes. The eyes of the horse fly, shown here, are very large and iridescent.

banded patterns caused by refraction and reflection of light at eye's surface

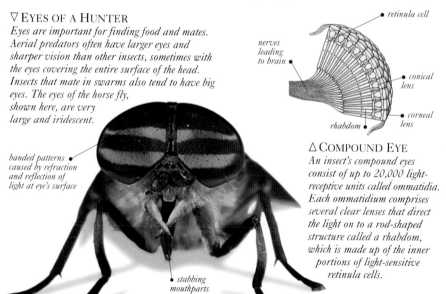

stabbing mouthparts

retinula cell

nerves leading to brain

conical lens

corneal lens

rhabdom

△ COMPOUND EYE

An insect's compound eyes consist of up to 20,000 light-receptive units called ommatidia. Each ommatidium comprises several clear lenses that direct the light on to a rod-shaped structure called a rhabdom, which is made up of the inner portions of light-sensitive retinula cells.

TOUCH, SMELL, AND TASTE

Chemical sense organs, or chemoreceptors, are present on the mouthparts, antennae, tarsi, and other parts of the body in insects. These enable the insect to detect food, find good egg-laying sites, or to follow marked trails on the ground. Insects pick up airborne odours by means of olfactory sensilla, which are located mainly on the antennae; if present in very large numbers, they can detect extremely low odour concentrations. Insects emit volatile chemicals called pheromones. These can be used for a variety of purposes, but are usually involved with sexual behaviour. Attraction pheromones act at a distance to bring the two sexes together; often it is the female who emits the odours and waits for males to find her. Once together, other odours called courtship pheromones are produced.

antennae used to communicate with other ants

COMMUNICATING ANTS
Messages are passed between insects in several ways. Some use sound or light displays, but touch and taste are more common. Here, two worker ants may be exchanging information about which colony they belong to and perhaps about new food sources.

DETECTING SOUND

Many insects have hairs on their cuticle surface that are responsive to vibrations, air currents, touch, and sound waves. Special hearing structures, called tympanal organs, may be present on various parts of the body (legs, wings, abdomen, or antennae). Depending on the species, these organs are responsive to sound frequencies ranging from well under 100Hz (cycles per second) to more than 200KHz. Male cicadas produce very loud sounds, which can be heard up to 1km (0.6 miles) away; the tympanal organs in both sexes are located in the abdomen.

Insects may use sound for a number of reasons: for attracting and finding a mate, detecting prey, and avoiding enemies. Many moths, preying mantids, lacewings, and several other species have ultrasonic-sensitive hearing organs, which allow them to receive the sounds of hunting bats.

POSITION OF THE "EARS"
Hearing organs are found on various parts of insects' bodies. In crickets, they are located on the tibiae on the front legs. Since the body of a cricket is fairly small, this gives better directional capability. The tympanal organ on each leg lies below two slits and is connected to special acoustic tracheae, which run back to the thorax.

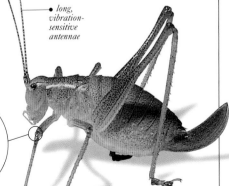

long, vibration-sensitive antennae

"ear" located on leg

FOOD AND FEEDING

ARTHROPODS EAT a variety of foods. Sometimes the food eaten by the immature stages and the adults is the same, but often the larval stages have very different feeding habits. In some cases, adult insects may not feed at all and simply depend on reserves built up at the larval stage. The main feeding types, together with the symbols that appear in this book, are outlined below.

PREDACIOUS SPECIES

Predacious species kill and eat other animals to survive. Most predators rely on more than one type of prey, although some do specialize. Predators do not have to eat as much as herbivores, since their food is more nutritious and provides all the protein that they need. Sometimes adult arthropods catch prey and store it for their larvae to eat. As a result of attempting to avoid predators, many arthropods have evolved defence mechanisms, including spines and hairs, cryptic colouring, and toxic secretions.

PRAYING MANTID ▷

Binocular vision allows the praying mantid to calculate the exact distance to its prey. The strike itself takes less than 100 milliseconds. The tibiae extend, then the femora, while the tibiae flex around the prey.

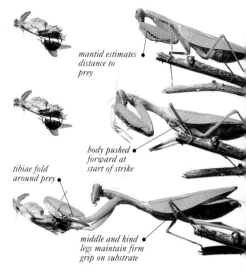

mantid estimates distance to prey

body pushed forward at start of strike

tibiae fold around prey

middle and hind legs maintain firm grip on substrate

HERBIVOROUS SPECIES

Plant-eating insects may feed on flowers, seeds, or leaves, or may eat inside the plants' tissues. A special case is that of gall-formers, which chemically induce an abnormal growth (gall) to form on a plant, inside which the insect is protected and feeds. Many insects have sucking mouthparts and feed only on plant sap or empty the contents of plant cells.

CATERPILLAR

One of the most well-known foliage-eaters, caterpillars use their thoracic legs and abdominal prolegs to grip leaves.

• caterpillar uses mandibles to nibble foliage

SAPROPHAGOUS SPECIES

Scavenging species that feed on dead or decaying organic matter are also called detritivores. Some of these scavengers eat primarily plant debris, while others devour mainly animal remains. In practice, it is difficult to distinguish precisely who eats what, and few species rely entirely on one type of food; for this reason, all scavenging species have been classified as saprophagous in this book.

FUNGIVOROUS SPECIES

Fungivorous species are those that are adapted to feeding on fungi (the fruiting body and the hidden hyphae). Typical examples are springtails and the larvae of many beetles and flies, which can be found inside the tissue of fungal fruiting bodies. Leaf-cutter ants and some species of termite cultivate fungal cultures for food.

HAEMATOPHAGOUS SPECIES

Ticks, fleas, many flies, and certain bugs need the blood of vertebrates to survive or bring their eggs to maturity. Some insects take only mammalian blood, whereas others feed on different hosts, such as birds or reptiles. Irritation from insect bites leads to scratching and sometimes to serious infections, even death. However, the main danger from bites lies in the transmission of various human and animal diseases caused by micro-organisms and protozoa. Malaria, yellow fever, and river blindness affect many millions of people in tropical regions; in temperate regions, ticks are significant disease vectors.

PARASITOIDS AND PARASITES

Parasitoids are specialized predators that live in or on the body of a host animal. In its lifetime, a parasitoid colonizes only one host, which it eventually kills. Some parasitoids feed internally (endoparasitoid), while others feed externally on the host's body (ectoparasitoid). Examples of this feeding strategy are found in parasitic wasps and some flies. For the purposes of this book, the same symbol is also given to parasites, such as fleas and lice, which feed on another animal's blood, skin, or hair, but do not kill their host.

INFESTED TREE HOPPER
This tree hopper (Membracidae, see p.98) is infested with red parasitic mites. The mites feed on the bugs' haemolymph by penetrating the cuticle with their mouthparts, especially at joints and where the cuticle is thin.

XYLOPHAGOUS SPECIES

Wood-eating species make use of an abundant resource, but it is of poor nutritional value, so many xylophages tend to be slow-growing. Some attack living or recently dead wood, while others can eat decaying wood only. Many have internal symbiotic micro-organisms to help them digest the cellulose; they may augment their diet by eating fungal hyphae and other material.

DEATHWATCH BEETLE
This pest lays its eggs in crevices on wood. The larvae burrow through the timber, and may take a few years to reach maturity.

larval feeding tunnels damage timber

COPROPHAGOUS SPECIES

Some species live on the droppings of other animals. Scarab beetles eat only dung, and in Africa, where there are numerous grazing mammals, there are thousands of species of dung beetle all using the droppings of various animals in different ways to rear their young. The larvae of many flies also breed in dung.

MELLIPHAGOUS SPECIES

With the exception of wind-pollinated grasses, most flowering plants are dependent on insects for pollination. In order to attract the right species, plants offer rewards of sugar-rich nectar and protein-rich pollen. While the insects supply their larval cells with these foods, the flowers are, in turn, pollinated.

PSEUDOPLACENTAL SPECIES

In some flies, such as the Tsetse Fly (see p.147), the larvae do not feed independently of the mother. The egg hatches and the larva is kept inside a brood chamber, where it feeds on secretions. When the larva is ready to be released, it may fill the whole of its mother's abdomen, and will usually pupate immediately.

ARTHROPOD BEHAVIOUR

THE BRAINS OF ARTHROPODS are relatively small. An adult locust, for example, has approximately one million nerve cells to serve all its sensory and motor needs. Smaller insects have far fewer nerve cells. Nevertheless, insects are capable of surprisingly sophisticated behaviour, which is evident in the way that they move, avoid predators, feed, mate, and care for their offspring.

COURTSHIP AND MATING

In most species of arthropod, the male and female need to mate before the female can lay her eggs. The sexes may simply meet at good feeding or egg-laying sites, or they may take a more active part in finding a suitable mate, attracting each other with songs, odours, and even light displays. At close range, courtship can be a complicated process. Insects may move their wings, legs, and antennae in certain ways, secrete pheromones, and give and receive nuptial gifts (usually pieces of food). Not all species have to mate, however. The females of many arthropod species are able to lay viable eggs without the need for males.

enlarged, toothed mandibles fit around rival's thorax

△ FIGHTING STAG BEETLES

Female arthropods are often highly selective, which leads to competition and rivalry between males. Here, male stag beetles fight for access to females. The winner may throw the opponent on to its back, and then mate with the female.

cold light, produced by a chemical reaction, shines through transparent cuticle

△ LIGHT ATTRACTION

In some beetles, such as the Glowworm, Lampyris noctiluca (above), females attract males of the same species by emitting flashes of light. In a few cases, females lure males of other species to eat them.

▷ GETTING TOGETHER

In many arthropods, sperm is transferred to the female indirectly. However, in insects (here, soldier beetles), copulation always takes place.

male about to pump sperm into female's sperm storage organ

CARE OF THE YOUNG

Parental care of eggs and young is fairly common in centipedes, arachnids, and insects. Female spiders wrap their eggs in silk and carry them around, or stay close by until they hatch. Scorpions and related groups brood their eggs and later carry their young on their backs. Among insects, it is usually the female who takes responsibility for "child-care", but the males may also play a part in some families.

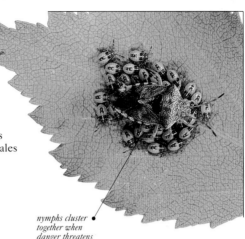

STANDING GUARD
Several species of bugs guard their young nymphs, and even guide them to good feeding sites. Here, a female parent bug watches over her brood resting on a birch leaf.

nymphs cluster together when danger threatens

METHODS OF SELF-DEFENCE

An arthropod's first line of defence is its cuticle, which may be very tough or leathery. Sharp, cuticular spines and protrusions, such as warts and bumps, or the ability to roll up into a ball, may further increase the protection that the exoskeleton provides. Mandibles and limbs are effective when used to strike out at enemies – the kick from a locust's hind leg can draw blood in most predators.

Physical defences are enhanced by producing unpleasant sounds, or repellent chemicals or odours. Many bugs, for instance, produce strong-smelling compounds from thoracic stink glands. Sap-sucking bugs, such as aphids, often surround themselves with "bodyguards" in the form of ants: the ants are attracted to the sugar-rich honeydew (excrement) that the bugs produce, and in turn help to protect the aphids from predators. Some arthropods are brightly coloured, which may serve to warn predators of their toxicity; sometimes eyespots and other bright patches are flashed at predators to startle them.

spines and hairs deter predators

△ POISONOUS PRICKLES
Toxic chemicals may be made inside the body or obtained from a poisonous food plant. These chemicals are often stored in outer parts of the body.

coloration of tiger moth indicates its unpalatability

leaf-like extensions

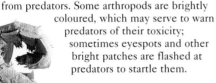

◁ CAMOUFLAGE
Many insects have evolved to blend into their surroundings, or to mimic dead leaves, sticks, thorns, bird droppings, stones, or even other, more dangerous, species.

△ WARNING COLOURS
Bright, contrasting warning colours advertise the presence of chemical defences. Some species "cheat", and are not actually poisonous.

SOCIAL INSECTS

MOST ARTHROPODS lead solitary lives, coming together only for mating. Some might be considered gregarious, grouping together for safety or sharing a food source. However, truly "social" species, (all termites and ants, some wasps and bees) are characterized by co-operation within a colony to rear young, coupled with a division of labour, and an overlap of generations.

SOCIAL WASPS AND BEES
In these highly social insects, the reproductive females, or queens, found and head the colonies. The queens lay eggs, then rear a few workers (sterile females) themselves. Thereafter, the queen leaves nest-building, colony defence, and feeding and tending the young to the growing number of workers. Mated queens can determine the sex of their offspring by withholding sperm if a male is preferred (males are produced from unfertilized eggs, females from fertilized ones).

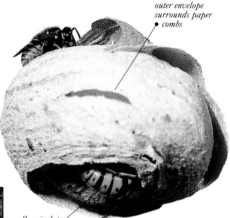

outer envelope surrounds paper combs

yellow-jacket worker extends the nest

◁ HONEYCOMB
The honey bee's nest consists of vertical wax combs divided into hexagonal cells, in which young are reared and honey is stored.

△ WASPS' NEST
Yellow jackets and hornets make exposed or underground nests of paper made from chewed wood fibres. The horizontal cells contain the developing larvae.

MIGRATING INSECTS

Some arthropods undertake regular migrations from one place to another to find food or egg-laying sites. Army ants (see p.184) and swarming species, such as the migratory locust *Locusta migratoria* (Acrididae, see p.64), are good examples of migratory insects. The longest insect migration is that of the Painted Lady butterfly, *Cynthia cardui* (Nymphalidae, see p.174), which can travel approximately 6,440km (4,000 miles) from North Africa to Iceland. Some spiders can be blown for hundreds of kilometres in wind currents.

LONG-DISTANCE TRAVELLERS
The Monarch butterfly (Danaus plexippus), a notable migrant, travels from winter roosts in Mexico to North America and Canada.

ANTS

These highly social insects,
which belong to the large family
Formicidae, are very abundant and have a
great impact on terrestrial ecosystems. In
most habitats, they are the major predators.

ants cooperate to carry leaf

Ants live in colonies ranging from a handful
of individuals to tens of millions. They
usually have female (queen), male, and
worker castes. The workers are all wingless,
sterile females; the larger ones may function
as soldiers to defend the colony. Reproduction
usually occurs between winged males and
females. After mating, the males usually die
and the females lose their wings. The caste
is mainly determined by the food that the
larvae are fed by the workers: a diet low in
protein will lead to the production of another
worker, whereas a diet high in protein will
produce a queen. The soldier's head and
jaws are often modified according to the
caste and species, and may be specialized for
crushing seeds or dismembering enemies.

△▽ LEAFCUTTER ANTS

*In Central and South America,
leafcutter ants (Atta species) are major
herbivores and among the most serious insect
pests. Their subterranean nests can be more than
5m (15ft) deep and comprise millions of workers.*

TERMITES

Termites, unlike many other insects, are able to
digest cellulose. In some tropical regions, they
are highly abundant and destructive, and may
eat up to one-third of the annual production of
dead wood, leaves, and grass. Termites live in
permanent social colonies and have a number
of distinct castes. The colonies normally have
a single queen, a king, and a few other
reproductive males. There may
also be supplementary
reproductives, which will
become active if anything should
happen to the queen. Termite
soldiers, unlike ants, are sterile males
and females. Worker termites resemble
the nymphs, and are the most numerous
caste. The role of the worker
termites is to build and repair
the nest, forage for food, and
feed the young nymphs.

central air shaft

food store

egg-production chamber

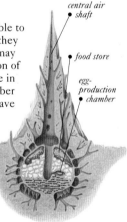

INSIDE THE NEST

*The internal structure of
many of the larger termite
nests allows air to circulate
so that the temperature
inside the nest can be
regulated to within 1°C
(1.8°F). Stale, carbon-
dioxide-rich air is vented to
the outside.*

TERMITE NEST

*Depending on the species, a
termite colony can range from a
tiny nest to a vast structure both
below and above ground.*

HABITATS

INSECTS AND OTHER terrestrial arthropods are found all over the globe, from snow-covered mountains to hot desert valleys, but they are not evenly distributed. Apart from some species of mites and midges in the Antarctic and some blood-sucking insects, such as mosquitoes, in the Arctic, there are very few arthropods near the poles. The closer to the equator you go, the greater the number of arthropods, both in species variety and abundance.

SURVIVAL STRATEGIES

The survival and persistence of most arthropod species has much to do with their relatively small size, protective cuticle, and ability to reproduce quickly (see p.14), but many also have special strategies for survival. When conditions are too hot and dry, many species remain dormant, while other species hibernate where winters are cold. Several insects, notably some species of ants, are able to function in extremely high temperatures, in excess of 65°C (149°F). At the other end of the scale, some species are able to withstand excessively cold conditions, surviving temperatures as low as - 40°C (- 40°F).

CONSERVATION

Mammals and birds used to be the only animals considered worthy of conservation, but a growing awareness of the vital role that insects play in global ecosystems is changing this view. Some rare insects are now protected by international law, and many countries are beginning to implement legislation. First and foremost, we must protect their habitats from destruction – the species will then look after themselves.

MOUNTAINS
Species that live in mountains are adapted to cold, wet, and windy conditions. Plant life becomes sparser with increasing altitude; as a result, there are fewer species of arthropod.

COMMON STONEFLY

TROPICAL FORESTS
These lush, moist habitats cover a tiny part of the total land area of the globe (about six per cent), yet they are estimated to hold approximately 50 per cent of all the world's arthropod species.

PRIAM'S BIRDWING

TEMPERATE WOODLAND
Although less lush than tropical forest, temperate woodland has a rich and varied fauna. Fertile soil, broad-leaved trees, deep leaf-litter, and decaying wood all provide ideal conditions for arthropods.

FUNNEL WEAVER

TEMPERATE GRASSLAND

Numerous insect species live in temperate grass-land. If heavily used for grazing or cultivation, insect diversity declines.

PLANT BUG

SAVANNAH

The canopies of scattered trees in tropical grassland harbour a rich, diverse arthropod fauna, particularly termites and ants. Overgrazing is endangering many species.

BRACONID WASP

CAVES AND DESERTS

Cave-dwelling species adapt to survive in total darkness and high humidity. Many are blind and wingless. Desert-dwellers adapt to temperature extremes and arid conditions.

BOTHRIURID SCORPION

FRESH WATER

Freshwater habitats have a unique arthropod fauna. Only about 5 per cent of insect species are aquatic for part of their lifecycle, yet their abundance means they contribute greatly to aquatic food chains.

WATER BOATMAN

SEASHORES

There are many opportunities for insects among the rocks, sand plants, and decaying seaweed along coasts. Beetles and flies are very abundant. A few insects are even adapted for going underwater occasionally.

PILL WOODLOUSE

TOWNS AND GARDENS

Many arthropods thrive in towns. Some species have taken up residence inside buildings or are associated with rubbish. The garden can be a wildlife refuge and home to numerous arthropod species.

BUMBLE-BEE

STUDYING INSECTS

YOU CAN LEARN a lot about insects from reference books, but if you really want to understand their world then you have to experience them at first hand. There is no substitute for patient observation: if you want to know exactly how a spider spins its web, it is best simply to sit and watch. Insects are endlessly fascinating, so take the time to observe their busy lives.

WHAT YOU WILL NEED

It is not necessary to invest in expensive equipment to study insects. Various sorts of collecting net are easy to make (see facing page). Below are a few other items that you will need, or are simply useful:

• Hand lens (X 10 magnification) and low-power lens, for viewing at close range.
• Camera, sketch book, and note book, to record your observations.
• Measuring tape and a stop watch, for determining running or flying speeds.
• Sieve, for sifting leaf-litter.
• Pooter (see facing page), also known as an aspirator, essential for collecting small insects without harming them.
• Pond net, a pair of wellingtons, and a few plastic containers for viewing aquatic species.

X 10 hand lens, essential for identification

low-power lens, for studying behaviour close up

IDENTIFYING INSECTS

Identifying an insect to the level of order is fairly straightforward with a little practice. However, to identify the species is considerably more difficult. While some insects are very distinctive, many bear an extremely close resemblance to other species, and are distinguishable only by minute characteristics that can be observed using a hand lens.

SLR camera with shoulder strap

macro lens for photographing objects close up

flowers are good places to photograph insects

PHOTOGRAPHY

A single lens reflex camera (SLR) with a 50mm or 100mm macro lens is ideal. Most zoom lenses with a macro setting are not good enough for photographing small insects. You may also need flash to maximize the depth of field.

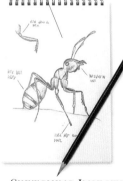

field notebook with blank pages for sketches and lined pages for notes

SKETCHING INSECTS

Drawing is an excellent way to record what you see, and you will also learn a great deal from close observation. Remember to record details in the book of when and where you found the insect.

HOW TO COLLECT INSECTS

Since most insects are mobile, collecting them involves intercepting them with hand nets, sweep nets, pond nets, or pooters as they fly, swim, or crawl. Trapping insects by attracting them to some kind of bait is also a good technique. In tropical regions, rotting fruit will attract many insects including butterflies, and animal dung will draw hundreds of beetles. Wherever you are, a piece of rotting fish in a plastic bottle will collect flies and some parasitic wasps.

If you want to discover how abundant a species is in a specific area compared to another site, it is important to carry out identical tests in each area for an accurate comparison. For example, make sure you cover the same surface area, use the same number of pitfall traps, or sieve the same volume of leaf-litter. Remember to wash your hands after field work.

USING A POOTER
With a pooter, you can collect small insects from a beating tray by sucking them up a tube and into a container.

BEATING TRAY
A white tray or cloth placed under a tree is a good way to catch insects: shake the branch to dislodge leaf-inhabiting species.

BUTTERFLY NET
These nets should be made of fine mesh and may be used to catch any flying insects. This one has an extendible handle.

PITFALL TRAP
To catch ground-running insects and other arthropods, sink a plastic cup into the ground with its top flush with the surface.

ATTRACTING INSECTS TO YOUR GARDEN

A garden with overgrown patches will attract considerably more forms of wildlife than one that is highly manicured. Variety is another important feature of a good wildlife garden: a wide range of habitats and microhabitats will attract a diverse fauna of insects and other animals. Never use pesticides in the garden.

When tidying the garden, try not to clear too much away that could be appealing to insects. Leave wood to decay naturally, and make a compost heap with vegetable waste from your kitchen and garden. A huge number of species live in decaying plant matter, and your garden will benefit from the compost you produce. Nectar-rich flowers such as lavender attract butterflies, moths, hover flies, and bees.

CREATE A WILDLIFE AREA
A pond surrounded by plants is ideal for luring wildlife to the garden, and provides a home and shelter for a great range of arthropods. Dig a pond as large as you can: aquatic insects will colonize it immediately.

IDENTIFICATION KEY

THIS KEY IS INTENDED to be a guide to the identification of specimens to the taxonomic level of order. First, answer the questions in keys 1 and 2. These will lead you to the catalogue of families and species that form the main part of the book – either directly or via the galleries of orders presented in keys 3–6. In most cases, close inspection with a hand lens will be necessary.

KEY 1: TERRESTRIAL ARTHROPODS

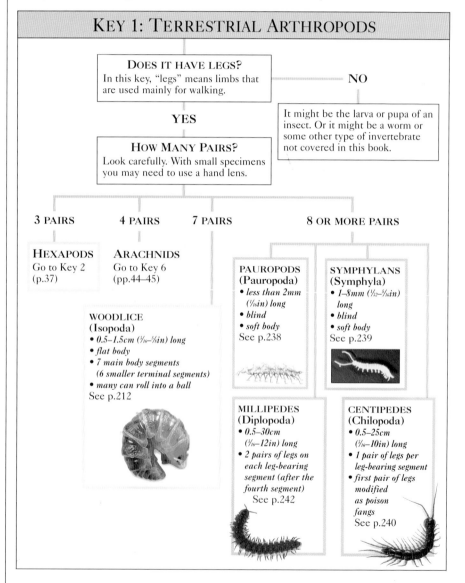

DOES IT HAVE LEGS?
In this key, "legs" means limbs that are used mainly for walking.

NO

It might be the larva or pupa of an insect. Or it might be a worm or some other type of invertebrate not covered in this book.

YES

HOW MANY PAIRS?
Look carefully. With small specimens you may need to use a hand lens.

3 PAIRS   **4 PAIRS**   **7 PAIRS**   **8 OR MORE PAIRS**

HEXAPODS
Go to Key 2
(p.37)

ARACHNIDS
Go to Key 6
(pp.44–45)

**WOODLICE
(Isopoda)**
• *0.5–1.5cm (³⁄₁₆–⁵⁄₈in) long*
• *flat body*
• *7 main body segments
(6 smaller terminal segments)*
• *many can roll into a ball*
See p.212

**PAUROPODS
(Pauropoda)**
• *less than 2mm
(⅟₁₆in) long*
• *blind*
• *soft body*
See p.238

**SYMPHYLANS
(Symphyla)**
• *1–8mm (⅟₃₂–⁵⁄₁₆in)
long*
• *blind*
• *soft body*
See p.239

**MILLIPEDES
(Diplopoda)**
• *0.5–30cm
(³⁄₁₆–12in) long*
• *2 pairs of legs on
each leg-bearing
segment (after the
fourth segment)*
See p.242

**CENTIPEDES
(Chilopoda)**
• *0.5–25cm
(³⁄₁₆–10in) long*
• *1 pair of legs per
leg-bearing segment*
• *first pair of legs
modified
as poison
fangs*
See p.240

KEY 2: HEXAPODS

DOES IT HAVE WINGS?
Look carefully, the wings may be very small or hidden under wing cases.

YES

WINGED INSECT
Go to Key 4
(pp.40–43)

NO

NON-INSECT HEXAPODS
These small, six-legged arthropods are all relatively primitive. They are distinguished from insects by their mouthparts.

IMMATURE INSECT
It might not be an adult specimen. In many orders, the immature stages are like small adults but without wings or genitalia.

WINGLESS INSECT
It may be a wingless adult insect. Go to Key 3
(pp.38–39)

SPRINGTAILS (Collembola)
- *less than 5mm (³/₁₆in) long*
- *often have a jumping organ (furcula) that can fold under the abdomen*
- *elongate or globular body shape*
See p.207

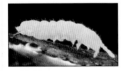

DIPLURANS (Diplura)
- *0.5–5cm (³/₁₆–2in) long*
- *blind*
- *pair of tail- or forcep-like appendages at end of abdomen*
See p.211

PROTURANS (Protura)
- *less than 2mm (¹/₁₆in) long*
- *blind*
- *pale*
- *soft body*
- *antennae very short or absent*
- *pointed head*
See p.210

KEY 3: WINGLESS INSECTS

Some insects are always wingless, notably fleas, lice, and the primitive silverfish and bristletails. However, many orders in which most insects have fully developed wings also contain species in which the wings are very short or absent. Winglessness is common in oceanic island and cave-dwelling species. Remember that a wingless specimen may be an immature winged insect.

ALWAYS WINGLESS

FLEAS
(Siphonaptera)
• *1–8mm (¹⁄₃₂–⁵⁄₁₆in) long*
• *brown with body flattened from side to side*
• *often found on animals or in nests*
• *can jump well*
See p.135

PARASITIC LICE
(Phthiraptera)
• *0.1–1cm (¹⁄₃₂–³⁄₈in) long*
• *flat body*
• *found on hair or feathers of host animals*
• *eyes small or absent*
• *legs modified to grip host*
See p.83

SILVERFISH
(Thysanura)
• *0.2–2cm (¹⁄₁₆–³⁄₄in) long*
• *three tail filaments*
• *abdominal segments with small ventral projections*
• *eyes small, not touching*
• *do not jump*
See p.47

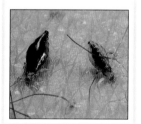

BRISTLETAILS
(Archaeognatha)
• *0.7–1.5cm (⁹⁄₃₂–⁵⁄₈in) long*
• *three tail filaments*
• *abdominal segments with small projections*
• *eyes touching*
• *humped body*
• *can jump*
See p.46

MAINLY WINGLESS

TERMITES
(Isoptera)
• *0.3–2cm (¹⁄₈–³⁄₄in) long*
• *pale body*
• *found in colonies*
See p.78

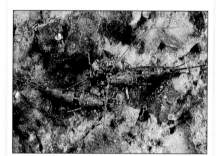

OCCASIONALLY WINGLESS

BEES, WASPS, & ANTS
(Hymenoptera)
- *0.25–70mm (¹/₂₈–2¾in) long*
- *often with constricted waist*
- *first abdominal segment fused to thorax*
- *many live in colonies*
- *often have sting*

See p.178

STICK INSECTS
(Phasmatodea)
- *1–30cm (⅜–12in) long*
- *stick-like body*
- *widely separated legs*

See p.66

BARKLICE & BOOKLICE
(Psocoptera)
- *1–9mm (¹/₂₂–¹¹/₃₂in) long*
- *soft, squat body*
- *humped back when seen from front*
- *large head and bulbous forehead*
- *bulging or reduced eyes*

See p.81

BUGS
(Hemiptera)
- *0.1–10mm (¹/₂₂–4in) long*
- *mouthparts form slender or short tube under head*
- *antennae have less than ten segments*

See p.85

SCORPIONFLIES
(Mecoptera)
- *0.3–3cm (⅛–1¼in) long*
- *head elongated downwards to form beak*

See p.133

THRIPS
(Thysanoptera)
- *0.5–12mm (¹/₆₄–½in) long*
- *slender, elongate body*
- *large faceted eyes*

See p.101

FLIES
(Diptera)
- *0.5–60mm (¹/₆₄–2½in) long*
- *very small prothorax*
- *vestigial wings (halteres) often present*

See p.136

MOTHS
(Lepidoptera)
- *body covered with scales*
- *proboscis usually coiled*

See p.158

KEY 4: WINGED INSECTS

The orders that appear here contain mainly winged insects although some of the groups do contain wingless species. Some beetles and bugs may not appear to have wings until examined closely. The wings are sometimes very small or hidden. As well as the more obvious details, such as the colour and shape of the wings, the way in which they are held also aids identification.

COMMON IN ALL HABITATS

BUTTERFLIES AND MOTHS
(Lepidoptera)
• *wingspan 0.3–30cm (⅛–12in)*
• *body and wings covered with scales*
• *proboscis often coiled*
• *long, thread-like antennae*
See p.158

FLIES
(Diptera)
• *0.5–60mm (¼4–2½in) long*
• *only one pair of wings (hindwings modified as halteres; sometimes hard to see)*
• *very small prothorax*
See p.136

NOTE
Look carefully: the forewings and hindwings might be joined by tiny hooks or hairs. If so, the specimen belongs to the Hymenoptera (see p.178).

BEETLES
(Coleoptera)
• *up to 18cm (7in) long*
• *forewings toughened to form elytra (wing cases)*
• *elytra may be short, leaving part of abdomen exposed*
See p.109

BEES, WASPS, & ANTS
(Hymenoptera)
• *0.25–70mm (½28–2⅜in) long*
• *often with constricted waist*
• *forewings wider or longer than hindwings*
• *wings joined in flight by tiny hooks*
• *many live in colonies*
See p.178

BUGS
(Hemiptera)
• *0.1–10cm (½2–4in) long*
• *mouthparts form a slender or short tube under the head*
• *front wings longer than hindwings*
• *stink glands may be present*
See p.85

NOTE
Look carefully: The mouthparts can be hard to see in some species.

MAINLY FOUND AROUND FRESH WATER

MAYFLIES
(Ephemoptera)
- *0.5–4cm (³/₁₆–1½in) long*
- *forewings large, triangular, and held upright*
- *abdomen with 2 or 3 filaments (tails)*
See p.48

CADDISFLIES
(Trichoptera)
- *0.2–4cm (¹/₁₆–1½in) long*
- *slender, moth-like body, covered with hairs*
- *long, thread-like antennae*
See p.156

STONEFLIES
(Plecoptera)
- *0.3–5cm (¹/₈–2in) long*
- *wings folded along body*
- *conspicuous cerci*
- *bulging eyes*
- *rectangular body*
See p.56

DAMSELFLIES & DRAGONFLIES
(Odonata)
- *4–15cm (1½–6in) long*
- *long, cylindrical abdomen*
- *large eyes*
- *wings of equal size*
- *thoracic segments slope backwards*
See p.51

ALDERFLIES & DOBSONFLIES
(Megaloptera)
- *1–15cm (³/₈–6in) long*
- *wings held together roof-like over body at rest*
- *long wings of similar size*
- *soft abdomen*
See p.103

THIS KEY IS CONTINUED OVER THE PAGE

KEY 4: WINGED INSECTS Continued

OTHER WINGED INSECTS

**TERMITES
(Isoptera)**
- *0.3–2cm (⅛–¾in) long*
- *wings with longitudinal veins and weak cross-veins*
- *only reproductives have wings, and these are shed after short nuptial flight*
- *very small cerci*
- *found in colonies*
See p.78

**SNAKEFLIES
(Raphidioptera)**
- *0.6–3cm (¼–1¼in) long*
- *wings held together roof-like over body at rest*
- *elongated cephalothorax*
See p.104

**LACEWINGS & ALLIES
(Neuroptera)**
- *0.2–9cm (⅟₁₆–3½in) long*
- *wings held together roof-like over body at rest*
- *net-like wing venation, with many cross-veins*
See p.105

**BARKLICE & BOOKLICE
(Psocoptera)**
- *1–9mm (⅟₃₂–¹¹⁄₃₂in) long*
- *soft, squat body*
- *humped back when seen from front*
- *large head and bulbous forehead*
- *bulging or reduced eyes*
See p.81

**GRASSHOPPERS &
CRICKETS
(Orthoptera)**
- *0.5–15cm (⅟₁₆–6in) long*
- *tough, leathery forewings*
- *hindlegs often large*
- *pronotum extended down at sides*
- *often produce sounds with wings or legs*
See p.60

**COCKROACHES
(Blattodea)**
- *0.3–10cm (⅛–4in) long*
- *flat, oval body*
- *shield-like pronotum, often covering head*
- *toughened forewings*
- *membranous hindwings*
See p.74

**STICK INSECTS
(Phasmatodea)**
- *1–30cm (⅜–12in) long*
- *stick- or leaf-like body*
- *widely separated legs*
See p.66

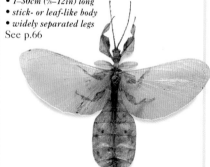

MANTIDS
(Mantodea)
- *0.8–15cm (⁵/₁₆–6in) long*
- *front legs distinctively modified for catching prey*
- *elongate prothorax*
- *toughened forewings*
- *large, membranous hindwings*

See p.71

EARWIGS
(Dermaptera)
- *0.5–5cm (³/₁₆–2in) long*
- *forceps at end of abdomen*
- *flat, elongate body*
- *short, tough forewings*
- *hindwings folded at rest*
- *telescopic and mobile abdomen*

See p.69

SCORPIONFLIES
(Mecoptera)
- *0.3–2.8cm (⅛–1⅛in) long*
- *head elongated downwards as a beak*
- *wings of similar size*

See p.133

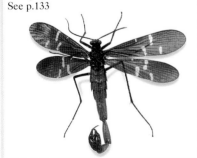

KEY 5: MINOR ORDERS

Some insects are very small and are unlikely to be encountered in the field. The females of certain species spend their entire lives hidden inside the body of a host insect.

WEB-SPINNERS
(Embioptera)
- *0.3–2cm (⅛–¾in) long*
- *swollen first tarsal segment of front legs*
- *live gregariously in silk tunnels*

See p.77

STREPSIPTERANS
(Strepsiptera)
- *up to 3.5cm (1½in) long*
- *females are wingless and live inside other insects*
- *males have fan-shaped hindwings and blackberry-like eyes*

See p.132

FEMALE

MALE

ANGEL INSECTS
(Zoraptera)
- *less than 5mm (³/₁₆in) long*
- *termite-like*
- *may be winged or wingless*

See p.80

ROCK CRAWLERS
(Grylloblattodea)
- *1.2–3cm (½–1¼in) long*
- *found in cold regions of western North America and eastern Asia*

See p.59

KEY 6: ARACHNIDS

Arachnids are wingless arthropods with four pairs of legs. Despite the fact that the group as a whole is very diverse in appearance, the major groups are easily recognized. Spiders are the most familiar arachnids. All spiders have silk-spinning organs, and some can be identified by the type of web they produce. Scorpions have distinctive claws on the end of their powerful pedipalps, and mobile abdominal "tails" with a sting. Pseudoscorpions, although quite similar to

SPIDERS
(Araneae)
• *up to 9cm (3½in) long*
• *relatively short pedipalps*
• *first pair of legs similar in size to other pairs*
• *non-segmented abdomen carries silk-spinning organ*

See p.228

WHIP-SPIDERS
(Amblypygi)
• *up to 4.5cm (1¾in) long*
• *squat body with broad cephalothorax*
• *large, spiny pedipalps*
• *first pair of legs very long*

See p.220

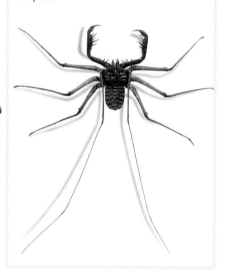

SUN-SPIDERS
(Solfugida)
• *up to 7cm (2¾in) long*
• *leg-like pedipalps*
• *large, forward-facing chelicerae*

See p.217

HARVESTMEN
(Opiliones)
• *up to 15cm (6in) long*
• *pedipalps have six segments*
• *legs usually long and slender*
• *segmented abdomen (sometimes hard to see)*

See p.221

scorpions, are much smaller and lack the "tail". Whip-scorpions have a distinctive whip-like "tail", and in whip-spiders the first pair of legs is very long and whip-like. Ticks and mites are distinctively rounded. Because of the dangerous nature of certain species, avoid handling an arachnid unless you are sure that it is a harmless specimen. In fact, the vast majority of arachnids are harmless. Scorpions and other arachnids will bite or sting human beings only in self-defence.

SCORPIONS (Scorpiones)
- *up to 18cm (7in) long*
- *flat body*
- *large pedipalps with pincers*
- *abdomen has long, jointed "tail" with a sting*

See p.213

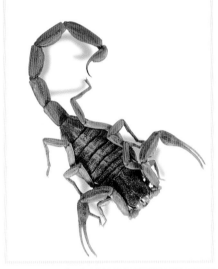

WHIP-SCORPIONS (Uropygi)
- *up to 7.5cm (3in) long*
- *powerful pedipalps*
- *abdomen has whip-like tail*

See p.219

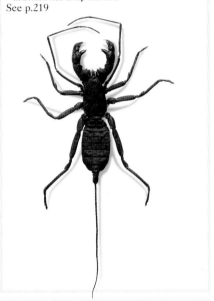

TICKS & MITES (Acari)
- *up to 3cm (1¼in) long*
- *body has no distinct divisions*
- *legs usually short*
- *non-segmented abdomen*

See p.223

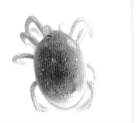

PSEUDOSCORPIONS (Pseudoscorpiones)
- *up to 1.2cm (½in) long*
- *flat body*
- *pedipalps like small scorpion claws*
- *oval abdomen with 11 or 12 segments*

See p.215

INSECTS

BRISTLETAILS

T HE ORDER ARCHAEOGNATHA contains 2 families and 350 species. These primitive, wingless insects look hump-backed when seen from the side. They have simple mouthparts, three ocelli, and large (compound) eyes that touch each other on top of the head. At the end of the abdomen are three long tails, the middle one being the longest. On the underside of the abdomen there are small projections, called styles, which help bristletails to move over steep surfaces more easily.

Bristletails are ametabolous: they continue moulting throughout their lives. Males deposit a sperm packet that is picked up by the female's genitalia. Small batches of eggs are laid in cracks and crevices, and the young may take up to two years to reach sexual maturity.

Order ARCHAEOGNATHA	Family MACHILIDAE	No. of species 250

JUMPING BRISTLETAILS

These elongate insects are covered with patterns of drab brown or dark grey scales. Many run rapidly and can also jump – hence the common name.
• LIFE-CYCLE Eggs are laid in small batches in cracks and crevices, and the young take about two years to reach adulthood. Jumping bristletails feed on lichen, algae, and plant debris.
• OCCURRENCE Worldwide. In grassland, in wooded or coastal areas, under stones, and in leaf-litter and decaying vegetable matter.

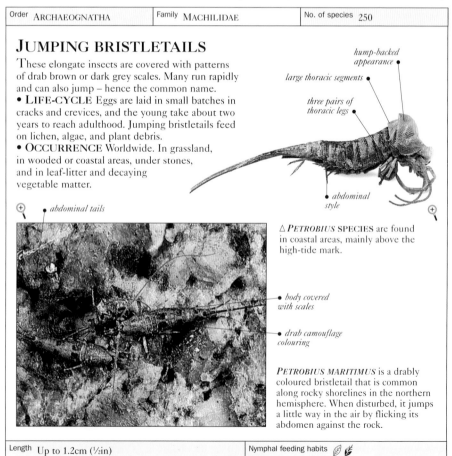

hump-backed appearance

large thoracic segments

three pairs of thoracic legs

abdominal style

abdominal tails

△ *PETROBIUS* SPECIES are found in coastal areas, mainly above the high-tide mark.

body covered with scales

drab camouflage colouring

PETROBIUS MARITIMUS is a drably coloured bristletail that is common along rocky shorelines in the northern hemisphere. When disturbed, it jumps a little way in the air by flicking its abdomen against the rock.

Length Up to 1.2cm (½in)	Nymphal feeding habits

SILVERFISH

T HE 4 FAMILIES AND 370 SPECIES of silverfish that make up the order Thysanura are primitive, wingless insects with elongate, flat bodies that may have scales on the surface. They have simple mouthparts and may have small, widely separated compound eyes or no eyes at all. Most species have no ocelli. The three abdominal tails are of equal length and, as in bristletails (see opposite page), the abdominal segments have projections called styles. Silverfish vary in appearance more than bristletails and occupy a wider range of habitats.

Males deposit sperm on silken threads, placed on the ground for females to pick up with their genitalia. Eggs are laid in cracks and crevices. Like bristletails, silverfish nymphs are ametabolous and develop without obvious metamorphosis.

Order THYSANURA	Family LEPISMATIDAE	No. of species 190

LEPISMATIDS

The brownish bodies of these insects are tapered, slightly flat, and usually covered with either greyish or silvery scales. They have compound eyes but no ocelli. All lepismatids are nocturnal. Some species favour cool, damp conditions, while others prefer warm, dry places.
• LIFE-CYCLE The females lay their eggs in cracks and crevices.
• OCCURRENCE Worldwide, especially in warmer regions. In tree canopies, under stones, and in caves; some inhabit houses or the nests of birds, ants, or termites.
• REMARK Domestic species eat flour, damp textiles, book bindings, and wallpaper paste.

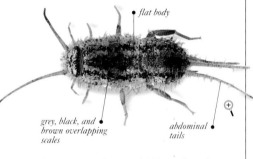

flat body

grey, black, and brown overlapping scales

abdominal tails

△ LEPISMATIDS in general (this specimen is an unidentified species) have brownish, scaly bodies. They reach sexual maturity after ten to twelve moults and may live for several years.

mottled scales

THERMOBIA DOMESTICA, or the Firebrat, is found all over the world. This species prefers warm habitats, such as areas in buildings near ovens and hot pipes. The antennae are as long as the body, and the longer hairs on the back of its body are arranged in groups on the rear margins of the body segments.

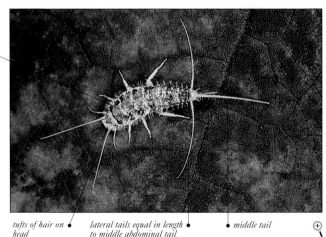

tufts of hair on head

lateral tails equal in length to middle abdominal tail

middle tail

Length 0.8–2cm (⁵⁄₁₆–¾in)	Nymphal feeding habits

MAYFLIES

T HE 23 FAMILIES and 2,500 species of the order Ephemeroptera – the mayflies – are the oldest, most primitive winged insects. They are also the only insects that moult after they have developed functional wings. Despite their common name, not all mayflies are common in May, and many species can be found at other times of the year.

Mayflies have soft bodies, long legs, and typically two pairs of wings. The forewings are large and triangular; the hindwings are small or may be absent altogether. Mayflies cannot fold their wings back along their body but instead hold them upwards or downwards.

Adults do not feed and live for a very short time – some species survive for just one day. Mating occurs in swarms at dawn or dusk, and females drop their eggs into water. Metamorphosis is incomplete. The aquatic nymphs, which usually have lateral abdominal gills and three terminal tails, eat a wide range of submerged plant and animal matter. When fully grown, they rise to the surface and moult into a form with dull-coloured wings, which is known as the sub-imago stage. They then leave the water. After a period of between one hour and several days, the final moult reveals the shiny-winged adult.

Order EPHEMEROPTERA	Family BAETIDAE	No. of species 900

SMALL MAYFLIES

These mayflies may be pale or dark brown or black with yellowish or grey markings. The forewings are elongate and rounded, and in some species the hindwings are either small or absent altogether.
• **LIFE-CYCLE** Eggs are dropped into water. The nymphs are either streamlined and swim well or slightly flat and crawl. They feed by scraping algae from surfaces.
• **OCCURRENCE** Worldwide. In streams, rivers, ponds, and lakes.
• **REMARK** Small mayflies can be found in higher, colder places than members of other mayfly families.

NYMPHS are small and streamlined in shape.

very small hindwings •   *two abdominal tails* •

yellow-tinged veins on forewing edge •   *brown* • *mottling*

△ **BAETIS RHODANI** is usually found in fast-flowing streams. The specimen shown here is the sub-imago stage, which has dull wings.

CLOEON DIPTERUM is a European species that breeds in a wide range of aquatic habitats, from ponds and ditches to water troughs and butts.

Length 0.3–1.4cm (⅛–⅝in), most 0.4–0.8cm (5/32–5/16in)	Nymphal feeding habits

| Order EPHEMEROPTERA | Family EPHEMERELLIDAE | No. of species 170 |

CRAWLING MAYFLIES

These medium-sized mayflies have three abdominal tails and drab or dark coloration.
• **LIFE-CYCLE** Many females release all their eggs in one clump, which separates out on the water's surface. The majority of nymphs feed on decaying organic matter.
• **OCCURRENCE** Worldwide, but rare in southern hemisphere. In rivers, streams, ponds, and lakes, among debris, weeds, and silt.
• **REMARK** Crawling mayflies are used as models for the artificial "flies" used in trout fly-fishing.

shiny wings

dull colouring

three abdominal tails

NYMPHS often have flat bodies.

EPHEMERELLA **SPECIES** are common in and around fast-flowing rivers and streams. A newly emerged adult specimen is shown here.

| Length 0.6–1.4cm (¼–⅝in) | Nymphal feeding habits |

| Order EPHEMEROPTERA | Family EPHEMERIDAE | No. of species 150 |

COMMON BURROWING MAYFLIES

The wings of these large mayflies are clear or brownish in colour, although they are dark-spotted in a few species. There are two or three long tails at the end of the abdomen.
• **LIFE-CYCLE** Females drop eggs into water. The nymphs dig and burrow into the silt at the bottom, aided by a process on the head and tooth-like mandibles. They mostly eat fine particles of organic material extracted from the silty bottom.
• **OCCURRENCE** Worldwide, except in Australia. In streams, rivers, and lakes.
• **REMARK** Some of the artificial "flies" used in trout fly-fishing are modelled on these mayflies.

large, richly veined, triangular forewings

thorax contains large flight muscles

small, rounded hindwings

three long abdominal tails

white or grey abdomen with dark markings

EPHEMERA DANICA is a large, widespread European species. It breeds in rivers and lakes with silty or sandy bottoms. Here, a female waits on a marginal plant before joining a mating swarm.

NYMPHS have large, tooth-like mandibles that help them to burrow.

| Length 1–3.4cm (⅜–1¼in) | Nymphal feeding habits |

Order EPHEMEROPTERA	Family HEPTAGENIIDAE	No. of species 500

STREAM MAYFLIES

These flat-headed mayflies are usually dark brown with clear wings and two long abdominal tails.
• LIFE-CYCLE Eggs are laid in water. The nymphs are typically active and live under stones and vegetation or in debris. Some are poor swimmers and cling tightly to rocks and stones on the bottom of the habitat. Most scrape algae or eat fine particles of organic matter.
• OCCURRENCE Worldwide, except Australia and New Zealand; rare in South America. In and around ponds, lakes, and fast-flowing streams.

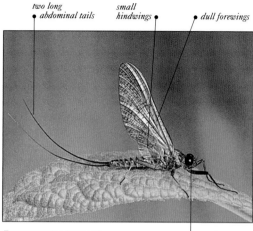

two long abdominal tails • small hindwings • dull forewings

• very large eyes in males

NYMPHS are flat and dark. They are highly mobile swimmers.

ECDYONURUS DISPAR is a common European mayfly. It favours lake shores and stony-bottomed rivers. The dull-winged sub-imago stage is seen here.

Length 0.4–1.5cm (⁵⁄₃₂–⁵⁄₈in), most 1cm (³⁄₈in)	Nymphal feeding habits

Order EPHEMEROPTERA	Family LEPTOPHLEBIIDAE	No. of species 600

PRONGILL MAYFLIES

Most of these drab mayflies have dark, longitudinal veins on the wings. The males' eyes are divided into an upward-facing region with large facets and a downward-facing region with smaller facets. The abdomen carries three long tails.
• LIFE-CYCLE Eggs are laid in water. The crawling nymphs may have a flat shape and live under stones and in debris. Most scrape algae or eat fine particles of organic debris; a few eat fish eggs.
• OCCURRENCE Worldwide. In streams and rivers; by the edges of ponds and lakes.
• REMARK The adults are used as models for artificial "flies" in trout fly-fishing.

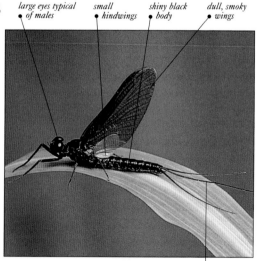

large eyes typical of males • small hindwings • shiny black body • dull, smoky wings

NYMPHS look grasshopper-like from the side. The abdomen has forked gills.

LEPTOPHLEBIA VESPERTINA is widespread in Europe, near lakes and small streams. Here, a sub-imago mayfly rests on vegetation before moulting into a shiny-winged adult.

• three abdominal tails

Length 0.4–1.4cm (⁵⁄₃₂–⁵⁄₈in), most 0.8–1cm (⁵⁄₁₆–³⁄₈in)	Nymphal feeding habits

DAMSELFLIES AND DRAGONFLIES

T HE 5,500 SPECIES and 30 families in the order Odonata are better known as damselflies and dragonflies. The damselflies are represented here by a selection of families, from the Calopterygidae to the Pseudostigmatidae. A selection of dragonflies follows, from the Aeshnidae to the Libellulidae.

The head of these insects has biting mouthparts, short antennae, and very large compound eyes. In damselflies, the head is broad, with widely spaced eyes, whereas dragonflies have rounded heads and eyes that are not widely separated. Both pairs of wings are more or less the same in damselflies, whereas the hindwings of dragonflies are broader than the forewings. At rest, damselflies fold their wings; dragonflies tend to hold them outstretched. Damselflies usually sit and wait for suitable prey, whereas dragonflies hunt prey in the air.

Males of both groups can curl their abdomen to transfer sperm from a genital opening on the ninth abdominal segment to a storage organ in the second or third. When mating, males may remove sperm from past matings. Eggs are laid in water and on aquatic plants. Metamorphosis is incomplete. The aquatic nymphs are predacious and have a hinged labium that can be shot forward to seize prey.

Order ODONATA	Family CALOPTERYGIDAE	No. of species 150

BROAD-WINGED DAMSELFLIES

These relatively large damselflies have wings that narrow gradually and appear to be unstalked. The wings may be dark and in males can have bright red marks at the bases or distinctive dark bands elsewhere. The pterostigma is small or may be absent altogether.

• LIFE-CYCLE Eggs are laid inside the tissues of various aquatic plants. Up to 300 eggs will be laid by a single female, and she may enter the water completely to do so. The nymphs hunt for prey in fast-flowing water.

• OCCURRENCE Worldwide, especially in warmer regions; rare in Australia. In fast- and slow-flowing rivers and streams, and canals.

• REMARK Adults may hunt some distance from water and prefer wooded areas.

NYMPHS have a small head and three prominent, flap-like gill filaments.

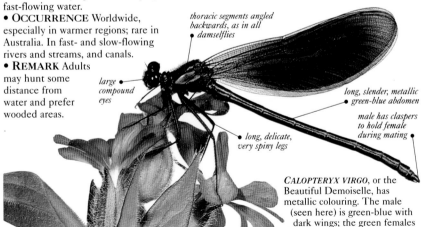

thoracic segments angled backwards, as in all damselflies

large compound eyes

long, slender, metallic green-blue abdomen

male has claspers to hold female during mating

long, delicate, very spiny legs

CALOPTERYX VIRGO, or the Beautiful Demoiselle, has metallic colouring. The male (seen here) is green-blue with dark wings; the green females have pale yellow-brown wings.

Wingspan 5–8cm (2–3¼in)	Nymphal feeding habits

Order ODONATA	Family COENAGRIONIDAE	No. of species 1,000

NARROW-WINGED DAMSELFLIES

Many of these slender damselflies are pale blue with dark markings. They rest horizontally with their relatively narrow wings folded above the body. Males are usually more brightly coloured than the females, which tend to be greyish or greenish.

• LIFE-CYCLE The female lays her eggs in water while the male is still grasping her by the neck. Using her egg-laying apparatus, she makes slits in the stems of aquatic plants and inserts small batches of eggs. The nymphs climb on to vegetation as they hunt for food.

• OCCURRENCE Worldwide, especially in temperate regions. In ponds, bogs, streams, and even brackish water.

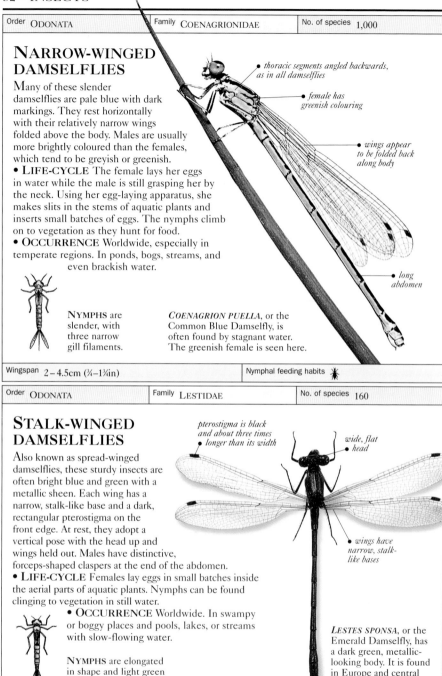

thoracic segments angled backwards, as in all damselflies

female has greenish colouring

wings appear to be folded back along body

long abdomen

NYMPHS are slender, with three narrow gill filaments.

COENAGRION PUELLA, or the Common Blue Damselfly, is often found by stagnant water. The greenish female is seen here.

Wingspan 2–4.5cm (¾–1¾in)	Nymphal feeding habits

Order ODONATA	Family LESTIDAE	No. of species 160

STALK-WINGED DAMSELFLIES

Also known as spread-winged damselflies, these sturdy insects are often bright blue and green with a metallic sheen. Each wing has a narrow, stalk-like base and a dark, rectangular pterostigma on the front edge. At rest, they adopt a vertical pose with the head up and wings held out. Males have distinctive, forceps-shaped claspers at the end of the abdomen.

• LIFE-CYCLE Females lay eggs in small batches inside the aerial parts of aquatic plants. Nymphs can be found clinging to vegetation in still water.

• OCCURRENCE Worldwide. In swampy or boggy places and pools, lakes, or streams with slow-flowing water.

NYMPHS are elongated in shape and light green to dark brown in colour.

pterostigma is black and about three times longer than its width

wide, flat head

wings have narrow, stalk-like bases

LESTES SPONSA, or the Emerald Damselfly, has a dark green, metallic-looking body. It is found in Europe and central and southern Asia.

Wingspan 3–7.5cm (1¼–3in)	Nymphal feeding habits ✷

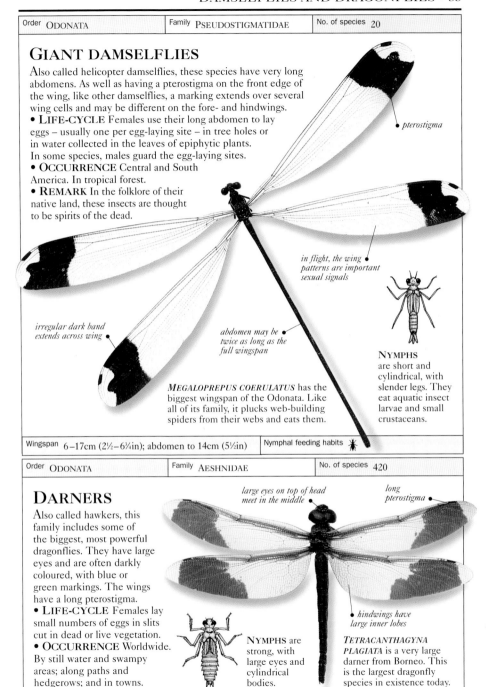

Order ODONATA	Family PSEUDOSTIGMATIDAE	No. of species 20

GIANT DAMSELFLIES

Also called helicopter damselflies, these species have very long abdomens. As well as having a pterostigma on the front edge of the wing, like other damselflies, a marking extends over several wing cells and may be different on the fore- and hindwings.
• LIFE-CYCLE Females use their long abdomen to lay eggs – usually one per egg-laying site – in tree holes or in water collected in the leaves of epiphytic plants. In some species, males guard the egg-laying sites.
• OCCURRENCE Central and South America. In tropical forest.
• REMARK In the folklore of their native land, these insects are thought to be spirits of the dead.

pterostigma

in flight, the wing patterns are important sexual signals

irregular dark band extends across wing

abdomen may be twice as long as the full wingspan

NYMPHS are short and cylindrical, with slender legs. They eat aquatic insect larvae and small crustaceans.

MEGALOPREPUS COERULATUS has the biggest wingspan of the Odonata. Like all of its family, it plucks web-building spiders from their webs and eats them.

Wingspan 6–17cm (2½–6¾in); abdomen to 14cm (5½in)	Nymphal feeding habits 🐜

Order ODONATA	Family AESHNIDAE	No. of species 420

DARNERS

Also called hawkers, this family includes some of the biggest, most powerful dragonflies. They have large eyes and are often darkly coloured, with blue or green markings. The wings have a long pterostigma.
• LIFE-CYCLE Females lay small numbers of eggs in slits cut in dead or live vegetation.
• OCCURRENCE Worldwide. By still water and swampy areas; along paths and hedgerows; and in towns.

large eyes on top of head meet in the middle

long pterostigma

hindwings have large inner lobes

NYMPHS are strong, with large eyes and cylindrical bodies.

TETRACANTHAGYNA PLAGIATA is a very large darner from Borneo. This is the largest dragonfly species in existence today.

Wingspan 6–14cm (2½–5½in), most 6.5–9cm (2½–3½in)	Nymphal feeding habits 🐜

Order ODONATA	Family CORDULEGASTRIDAE	No. of species 50

BIDDIES

These large dragonflies
are brownish or black
with yellow markings.
The eyes touch each other
at a single point. Most
members of this family
have a long abdomen.

• LIFE-CYCLE
Females lay their eggs at
the bottom of fast-flowing
streams and rivers. The nymphs live
buried in the mud or gravel, with just the head
and front legs exposed to seize passing prey.
Nymphs may take up to five years to mature, and
many species live for only a few weeks as adults.

• OCCURRENCE Northern hemisphere. Along
mountain or woodland streams; sometimes over
open ground near pools.

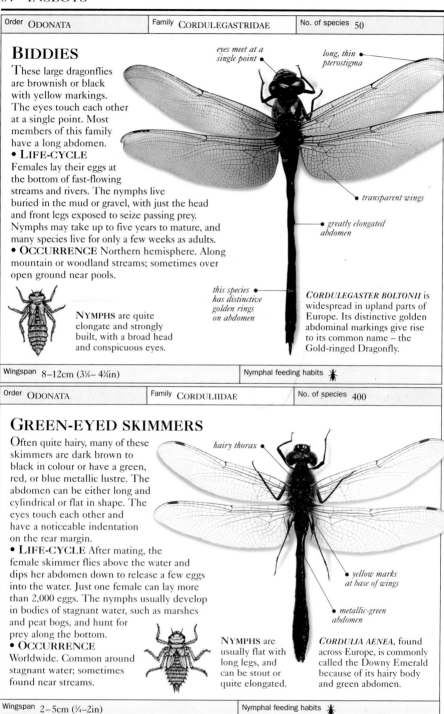

eyes meet at a single point

long, thin pterostigma

transparent wings

greatly elongated abdomen

NYMPHS are quite
elongate and strongly
built, with a broad head
and conspicuous eyes.

this species has distinctive golden rings on abdomen

CORDULEGASTER BOLTONII is
widespread in upland parts of
Europe. Its distinctive golden
abdominal markings give rise
to its common name – the
Gold-ringed Dragonfly.

Wingspan 8–12cm (3¼– 4¾in)	Nymphal feeding habits

Order ODONATA	Family CORDULIIDAE	No. of species 400

GREEN-EYED SKIMMERS

Often quite hairy, many of these
skimmers are dark brown to
black in colour or have a green,
red, or blue metallic lustre. The
abdomen can be either long and
cylindrical or flat in shape. The
eyes touch each other and
have a noticeable indentation
on the rear margin.

• LIFE-CYCLE After mating, the
female skimmer flies above the water and
dips her abdomen down to release a few eggs
into the water. Just one female can lay more
than 2,000 eggs. The nymphs usually develop
in bodies of stagnant water, such as marshes
and peat bogs, and hunt for
prey along the bottom.

• OCCURRENCE
Worldwide. Common around
stagnant water; sometimes
found near streams.

hairy thorax

yellow marks at base of wings

metallic-green abdomen

NYMPHS are
usually flat with
long legs, and
can be stout or
quite elongated.

CORDULIA AENEA, found
across Europe, is commonly
called the Downy Emerald
because of its hairy body
and green abdomen.

Wingspan 2–5cm (¾–2in)	Nymphal feeding habits

| Order ODONATA | Family GOMPHIDAE | No. of species 950 |

CLUB-TAILED DRAGONFLIES

The common name of these relatively large dragonflies refers to the unusually shaped abdomen of the males – and often of the females. The abdomen is swollen just before its apex, giving it the appearance of a club. Club-tailed dragonflies have widely separated eyes, and most species are brightly coloured in differing combinations of black, yellow, or green.

• LIFE-CYCLE Mating takes place among vegetation. Eggs are laid in shallow water, with the female lashing the surface with the end of the abdomen in order to release the eggs. The nymphs are bottom-living, crawling and burrowing to catch their prey.

• OCCURRENCE Worldwide. In ponds, lakes, rivers, and streams.

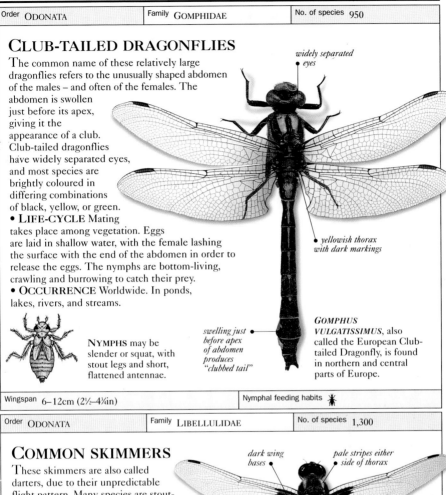

widely separated eyes

yellowish thorax with dark markings

NYMPHS may be slender or squat, with stout legs and short, flattened antennae.

swelling just before apex of abdomen produces "clubbed tail"

GOMPHUS VULGATISSIMUS, also called the European Club-tailed Dragonfly, is found in northern and central parts of Europe.

| Wingspan 6–12cm (2½–4¾in) | Nymphal feeding habits 🐜 |

| Order ODONATA | Family LIBELLULIDAE | No. of species 1,300 |

COMMON SKIMMERS

These skimmers are also called darters, due to their unpredictable flight pattern. Many species are stout-bodied, and they are often extremely colourful. The wings may have dark, irregular markings.

• LIFE-CYCLE Females usually hover over water and dip or strike their abdomen below the surface to release the eggs, which fall on to plants or to the bottom. The nymphs hunt for prey in mud and debris or on plants.

• OCCURRENCE Worldwide. In various habitats, including forests and mountainous areas, near slow-flowing streams, ponds, and bogs.

dark wing bases

pale stripes either side of thorax

distinct pterostigma

yellow marks at side of abdomen in both sexes

relatively broad, flat abdomen

male has pale blue abdomen (female's is brownish yellow)

NYMPHS are often squat and slightly flat in shape.

LIBELLULA DEPRESSA is a European species whose adults fly in June and July. Its nymphs take up to three years to mature.

| Wingspan 2–9.5cm (¾–3¾in) | Nymphal feeding habits 🐜 |

STONEFLIES

T HE 15 FAMILIES AND 2,000 species of stoneflies form the order Plecoptera. The body of a stonefly is typically soft, relatively flat, and slender. The elongate abdomen ends in a pair of tails (cerci), while the legs are sturdy. Although stoneflies have two pairs of wings, they are not strong fliers and are never found very far from water. The mouthparts are either under-developed or absent. In many species, the adults are short-lived and do not feed at all.

In the courting rituals of these insects, the males of many species attract females by drumming the underside of the abdomen against the ground or by trembling. In most species, the males and females "duet" – that is, they send courtship sounds back and forth between them. After mating on plant matter or on the ground, the females lay egg masses in water. Metamorphosis is incomplete. Most aquatic nymphs have gill-tufts and two terminal abdominal filaments. The nymphs pass through more than 30 moults before they finally emerge as adults.

Stoneflies occur all over the world but are most common in cool, temperate regions. However, there are five stonefly families that are found only in the southern hemisphere.

Order PLECOPTERA	Family CAPNIIDAE	No. of species 250

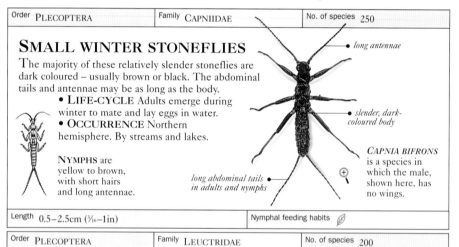

SMALL WINTER STONEFLIES

The majority of these relatively slender stoneflies are dark coloured – usually brown or black. The abdominal tails and antennae may be as long as the body.

• **LIFE-CYCLE** Adults emerge during winter to mate and lay eggs in water.
• **OCCURRENCE** Northern hemisphere. By streams and lakes.

NYMPHS are yellow to brown, with short hairs and long antennae.

long antennae

slender, dark-coloured body

CAPNIA BIFRONS is a species in which the male, shown here, has no wings.

long abdominal tails in adults and nymphs

Length 0.5–2.5cm (³⁄₁₆–1in)	Nymphal feeding habits

Order PLECOPTERA	Family LEUCTRIDAE	No. of species 200

ROLLED-WINGED STONEFLIES

At rest, the wings of these slender, usually dark brown stoneflies are either folded down or rolled around the sides of the body – hence the name. The cerci are very short.
• **LIFE-CYCLE** Adults often emerge very early in the year to mate and lay eggs in water.
• **OCCURRENCE** Mainly northern hemisphere. Often by small streams, springs, or lakes.

downward-pointing head

slender legs

NYMPHS are slender and yellowish in colour.

LEUCTRA SPECIES have dull colouring. At rest, their wings are rolled tightly around the body.

Length 0.6–1.3cm (¼–½in), most under 1cm (³⁄₈in)	Nymphal feeding habits

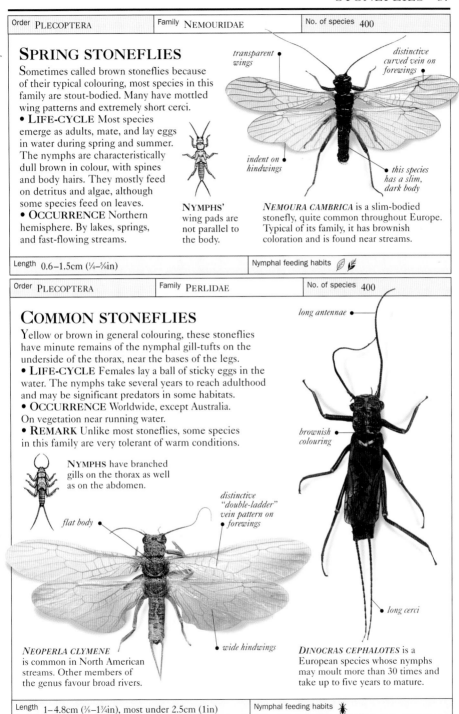

Order PLECOPTERA	Family NEMOURIDAE	No. of species 400

SPRING STONEFLIES

Sometimes called brown stoneflies because of their typical colouring, most species in this family are stout-bodied. Many have mottled wing patterns and extremely short cerci.
• **LIFE-CYCLE** Most species emerge as adults, mate, and lay eggs in water during spring and summer. The nymphs are characteristically dull brown in colour, with spines and body hairs. They mostly feed on detritus and algae, although some species feed on leaves.
• **OCCURRENCE** Northern hemisphere. By lakes, springs, and fast-flowing streams.

transparent wings

distinctive curved vein on forewings

indent on hindwings

this species has a slim, dark body

NYMPHS' wing pads are not parallel to the body.

NEMOURA CAMBRICA is a slim-bodied stonefly, quite common throughout Europe. Typical of its family, it has brownish coloration and is found near streams.

Length 0.6–1.5cm (¼–⅝in)	Nymphal feeding habits

Order PLECOPTERA	Family PERLIDAE	No. of species 400

COMMON STONEFLIES

Yellow or brown in general colouring, these stoneflies have minute remains of the nymphal gill-tufts on the underside of the thorax, near the bases of the legs.
• **LIFE-CYCLE** Females lay a ball of sticky eggs in the water. The nymphs take several years to reach adulthood and may be significant predators in some habitats.
• **OCCURRENCE** Worldwide, except Australia. On vegetation near running water.
• **REMARK** Unlike most stoneflies, some species in this family are very tolerant of warm conditions.

long antennae

brownish colouring

NYMPHS have branched gills on the thorax as well as on the abdomen.

distinctive "double-ladder" vein pattern on forewings

flat body

long cerci

NEOPERLA CLYMENE is common in North American streams. Other members of the genus favour broad rivers.

wide hindwings

DINOCRAS CEPHALOTES is a European species whose nymphs may moult more than 30 times and take up to five years to mature.

Length 1–4.8cm (⅜–1¾in), most under 2.5cm (1in)	Nymphal feeding habits

Order PLECOPTERA	Family PERLODIDAE	No. of species 250

PREDATORY STONEFLIES

Species in this family range from pale yellow to dark brown. They have a square pronotum and long abdominal tails.
• LIFE-CYCLE Eggs are generally laid in streams in spring. Fully grown nymphs crawl on to stones to emerge as adults in late spring or early summer. The adults are active during the day, rarely feed, and mostly die soon after laying their eggs. The nymphs are largely predacious, although they may eat plant matter or rotting material when very young.
• OCCURRENCE Northern hemisphere. In and around medium- to large-sized, stony-bottomed streams.

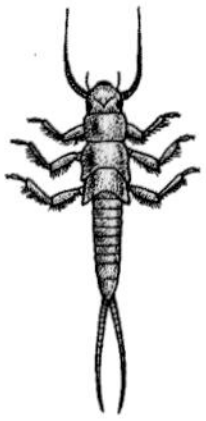

yellow-red central stripe on head and pronotum

brownish coloration

multi-segmented cerci

NYMPHS are waxy-looking, with light and dark patterning and long legs.

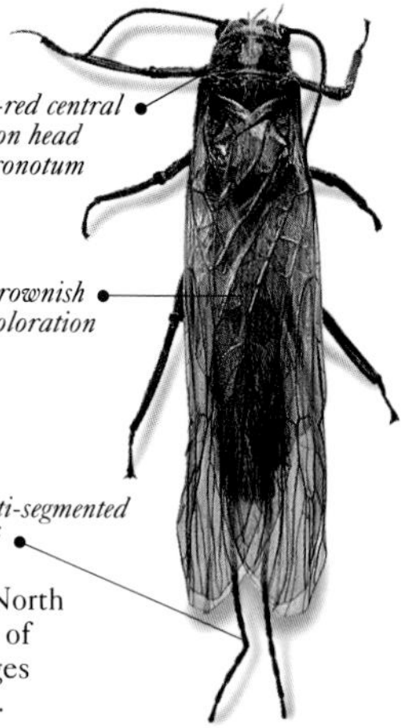

HYDROPERLA CROSBYI is a North American species. The sexes of this species exchange messages via special drumming sounds.

Length 0.8–5cm (⁵⁄₁₆–2in), most under 2.5cm (1in)	Nymphal feeding habits

Order PLECOPTERA	Family PTERONARCYIDAE	No. of species 12

GIANT STONEFLIES

The adults of this family have broad, heavy bodies. Their coloration is either grey or brown. Giant stoneflies are active mostly after dark.
• LIFE-CYCLE Dark, rounded eggs are laid in water. The nymphs take up to three years to develop, and use their jaws to shred aquatic plants and detritus. Adults emerge in summer and do not feed.
• OCCURRENCE Northern hemisphere. In streams and rivers.
• REMARK *Pteronarcys californica* is the well-known "salmonfly" (newly emerged adults are salmon-pink), used by anglers to catch trout in western North America.

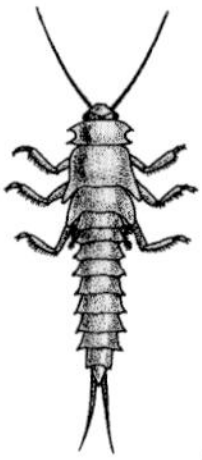

NYMPHS can be large and often have distinctive lateral expansions on the thorax and abdomen.

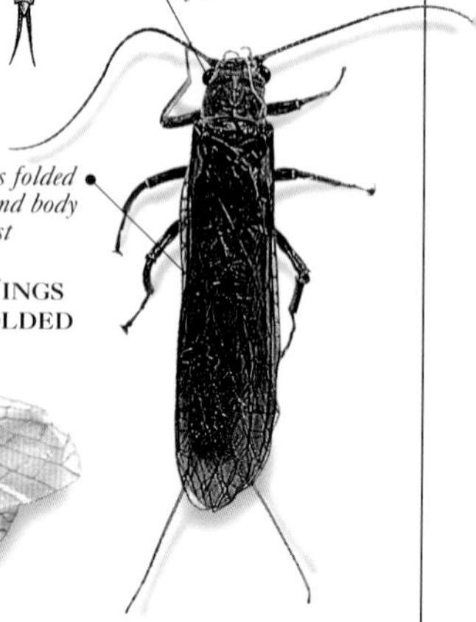

relatively large eyes

wings folded around body at rest

brown coloration

WINGS FOLDED

area of hindwing with no cross veins

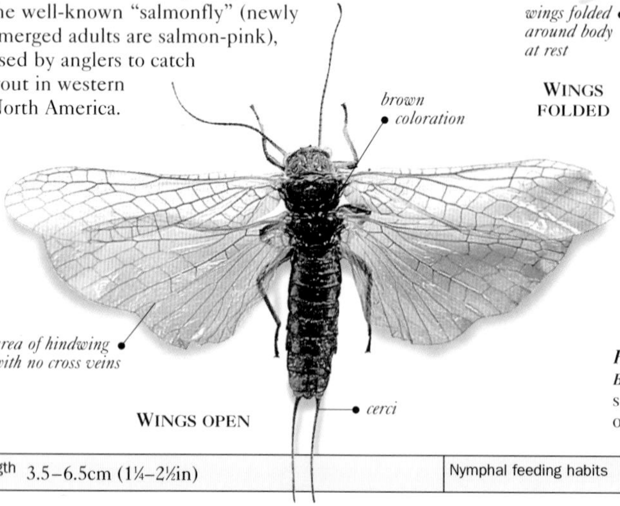

WINGS OPEN

cerci

PTERONARCELLA BADIA is a widespread species in western parts of North America.

Length 3.5–6.5cm (1¼–2½in)	Nymphal feeding habits

ROCK CRAWLERS

T HE ORDER GRYLLOBLATTODEA consists of a single family, divided into 25 species. These small and wingless insects were first discovered in the Canadian Rockies in 1906, and were initially considered to be a primitive family belonging to the order Orthoptera (see pp.60–65).

Rock crawlers have slender cerci at the end of their abdomen. The small head bears thread-like antennae with 22 to 40 segments and simple, biting, forward-facing mandibles. Eyes may be small or totally absent. Early stages look rather like immature earwigs, and the females have a short ovipositor.

Rock crawlers are found in eastern Asia and North America. Most species belong to the genus *Grylloblatta*, which is native to western United States and Canada. The members of this genus are adapted to mountainous conditions and low temperatures and are commonly found in rotting wood or moving over rocks, snow, and ice after dark – hence the common name, ice bugs.

Rock crawlers are good daytime and nocturnal hunters, eating live, recently dead, windblown, or torpid prey items. They may also eat moss and plant matter, especially when they are young. Metamorphosis is incomplete.

Order GRYLLOBLATTODEA	Family GRYLLOBLATTIDAE	No. of species 25

ROCK CRAWLERS

These insects are pale brown, yellow-brown, or grey. The body has a covering of short hairs, which may be dense or sparse, depending on species.

• **LIFE-CYCLE** Mating may last up to four hours, and the female may not lay her eggs for several months. When she does, she lays them in rotten wood, moss, rock crevices, and soil. Nymphal development may take more than five years, and there may be as many as nine nymphal stages.

• **OCCURRENCE** Cooler parts of the northern hemisphere. In sub-alpine deciduous forests, mountainous areas, and in limestone caves.

flat head

abdomen lighter in colour at sides

overall pale yellow coloration

cerci

pale antennae

pale, slender legs

ovipositor just a little shorter than cerci

△ *GRYLLOBLATTA CAMPODEIFORMIS*, or the Northern Rock Crawler, is found at high altitudes in North America and Canada, near glaciers and on rocks and damp scree slopes.

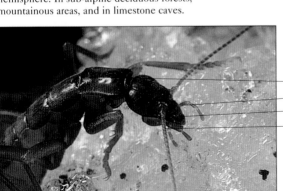

• *rectangular pronotum*

• *palps*

• *forward-facing mandibles*

• *antennae*

GRYLLOBLATTA SPECIES are found under stones and on open ground in autumn and spring. They spend the short summers underground or in crevices.

Length 1.2–3cm (½–1¼in)	Feeding habits 🐜 🦗

CRICKETS AND GRASSHOPPERS

T HE 28 FAMILIES AND 20,000 species of crickets, grasshoppers, and their relatives form the order Orthoptera. They have chewing mouthparts and hindlegs that are adapted for jumping. Most species have toughened forewings to protect the larger hindwings.

These insects are found in a range of terrestrial habitats. Singing is common, usually by males to attract mates, and metamorphosis is incomplete. There are two sub-orders: Ensifera and Caelifera. The Ensifera, typical of tropical and subtropical regions, comprise crickets and katydids (the selection here runs from Gryllacrididae to Tettigoniidae). The Caelifera, dominant in temperate areas, comprise grasshoppers and locusts (the selection of families here runs from Acrididae to Tetrigidae).

Order ORTHOPTERA	Family GRYLLACRIDIDAE	No. of species 600

LEAF-ROLLING CRICKETS

True to their name, many of these crickets roll leaves into a kind of nest in which they hide during the day. The antennae are long and thread-like. In females, the ovipositor is long – often longer than the rest of the body – and has a slight upward curve.
• LIFE-CYCLE Eggs are laid on bark, vegetation, and sometimes on the ground.
• OCCURRENCE Mainly tropical regions. In trees; sometimes on lower vegetation or the ground.

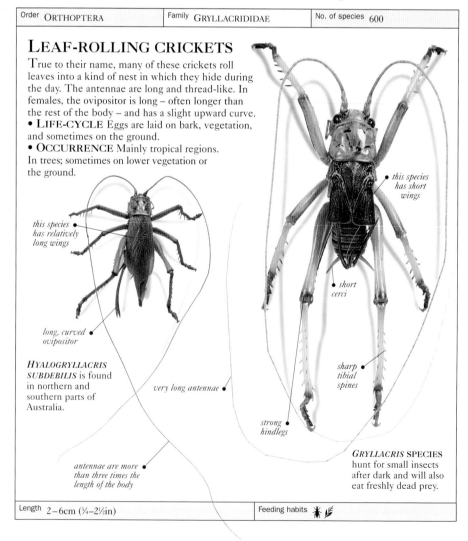

this species has short wings

this species has relatively long wings

short cerci

long, curved ovipositor

HYALOGRYLLACRIS SUBDEBILIS is found in northern and southern parts of Australia.

very long antennae

sharp tibial spines

strong hindlegs

antennae are more than three times the length of the body

GRYLLACRIS SPECIES hunt for small insects after dark and will also eat freshly dead prey.

Length 2–6cm (¾–2½in)	Feeding habits 🐜 🌿

Order ORTHOPTERA	Family GRYLLIDAE	No. of species 4,000

TRUE CRICKETS

These insects are slightly flat in shape, with rounded heads, and long, thread-like antennae. When present, the wings are folded flat over the body. Most are various shades of brown or black. Males sing mating songs using ridges at the bases of the forewings. In most females, the ovipositor is conspicuous and either cylindrical or needle-like.

• LIFE-CYCLE Eggs are laid in damp soil, singly or in masses. The carnivorous species known as tree crickets lay their eggs in small groups inside plant tissue. There are up to 12 nymphal stages.

• OCCURRENCE Worldwide. In woods, meadows, scrub, and grassland. Most species are ground-living.

• REMARK Some species are renowned for their songs, and in certain countries are kept in cages, as pets.

rounded head

brown coloration

wings folded flat over body

stout hindlegs

long, thread-like antennae

sensitive cerci

GRYLLUS BIMACULATA, the Two-spotted Cricket, is common in Africa, southern Europe, and Asia.

BRACHYTRUPES SPECIES include some damaging pests, which can attack the seedlings of valuable crops such as tea, tobacco, and cotton.

Length 0.5–5cm (³⁄₁₆–2in)	Feeding habits 🐜 🍃

Order ORTHOPTERA	Family GRYLLOTALPIDAE	No. of species 60

MOLE CRICKETS

These brownish, burrowing crickets – which are covered with short, velvety hairs and have short, broad front legs adapted for digging – look remarkably like miniature moles. They are stout with short, leathery forewings.

• LIFE-CYCLE Mating takes place on the surface, and eggs are laid in underground chambers. There are about ten nymphal stages, and the nymphs stay underground, eating plant roots and stems and small prey.

• OCCURRENCE Worldwide. In burrows up to 20cm (8in) long in damp sand or soil near streams, ponds, or lakes.

• REMARK Males produce songs by rubbing their forewings together. Their burrows may have flared tunnels that amplify and carry the song to the surface.

toothed front legs used for digging

hindlegs used to push soil back along tunnels

GRYLLOTALPA GRYLLOTALPA, commonly known as the European Mole Cricket, can be a pest of grasses, vegetables, and other crops. It is a protected species in the UK.

Length 2–4.5cm (¾–1¾in)	Feeding habits 🐜 🍃

| Order ORTHOPTERA | Family RHAPHIDOPHORIDAE | No. of species 500 |

CAVE CRICKETS

Many of these hump-backed, squat, wingless crickets have very long hindlegs and extremely long antennae, which are waved backwards and forwards in the dark to sense the environment and approaching predators. They are drably coloured, usually brown or grey. Some highly cave-adapted species have reduced eyes and soft bodies.
• **LIFE-CYCLE** Eggs are laid in the substrate of the cave, and the nymphs search for food as soon as they hatch. Nymphs of some species eat plant life at cave entrances.
• **OCCURRENCE** Widespread, especially in warmer regions. In caves and other humid locations, under stones and logs.

very long antennae for detecting predators

distinctive humped back

stout hind femur

long hindlegs adapted for jumping

DIESTRAMMENA MARMORATA has very long hindlegs with thick femora and distinctive mottling.

PHOLEOGRYLLUS GEERTSI is a species that is native to northern Africa and parts of southern Europe.

| Length 1.3–3.8cm (½–1½in) | Feeding habits 🦗 🌿 |

| Order ORTHOPTERA | Family STENOPELMATIDAE | No. of species 40 |

KING CRICKETS

Some members of this family are known as stone or Jerusalem crickets or as wetas. Most are large, chunky, wingless insects, with a large head and relatively short antennae. The legs are stout, and the tibiae have rows of strong spines, which are used for digging. Most species are very dark brown or black. Adults emerge from their underground burrows only after darkness falls.
• **LIFE-CYCLE** Eggs are laid in soil. King crickets may have nine or ten nymphal stages.
• **OCCURRENCE** Warm tropical regions. In rotten wood and underground.

stout legs

two rows of sharp tibial spines

hindlegs much bigger than other two pairs

short cerci

ovipositor

DEINACRIDA RUGOSA, probably once common across New Zealand, now survives only on islands where introduced animals, such as mice, are absent or controlled.

| Length 3.8–8cm (1½–3¼in) | Feeding habits 🦗 🌿 |

| Order ORTHOPTERA | Family TETTIGONIIDAE | No. of species 6,000 |

KATYDIDS

Also called bush crickets or long-horned grasshoppers, katydids are named after a species whose song sounds like the phrase "Kate-she-did". Most are large brown or green insects with wings that slope over the sides of their bodies. Many species mimic leaves, bark, or lichen, and some flash bright hindwing colours to startle predators.

• **LIFE-CYCLE** Male katydids sing to attract mates, using a file-and-scraper system at the base of the forewings. Females lay rows of eggs in plants or soil, and there may be five or six nymphal stages.

• **OCCURRENCE** Worldwide, mainly in tropical regions. On vegetation and in forests, from ground level to the canopy.

▷ *SATHROPHYLLIA RUGOSA*, from India, mimics the appearance of bark and is very difficult to see when at rest.

▽ *METRIOPTERA BRACHYPTERA*, also known as the Bog Bush Cricket, lives in damp meadows and is widespread across Europe.

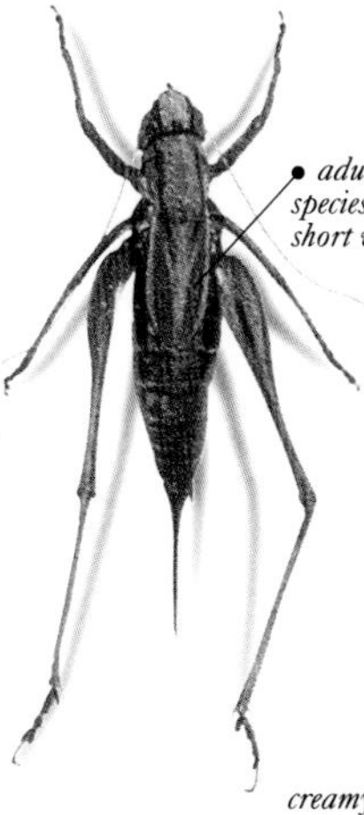

adults of this species have short wings

uneven outline of wings increases resemblance to leaves

slender, sickle-shaped ovipositor for laying eggs in plant tissue

very long, multi-segmented antennae

creamy brown coloration to blend with bark

▽ *LEPTOPHYES PUNCTATISSIMA*, or the Speckled Bush Cricket, is commonly found in bushes, trees, and grassland. The female has a large, sword-like ovipositor.

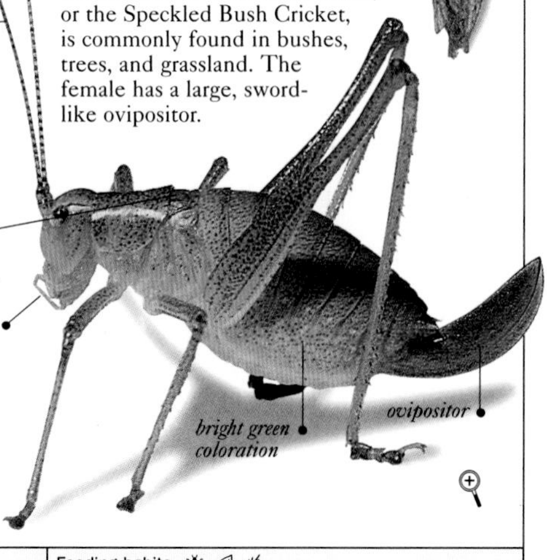

very short wings in both sexes

downward-pointing mouthparts

△ *OMMATOPTERA PICTIFOLIA* comes from Brazil. Like many bush crickets, it mimics dead leaves and keeps quite still when predators are close at hand.

bright green coloration

ovipositor

| Length 1.5–7.5cm (⅝–3in), most 3.5–5cm (1¼–2in) | Feeding habits 🐜 🍃 🌿 |

Order ORTHOPTERA	Family ACRIDIDAE	No. of species 10,000

GRASSHOPPERS

Most grasshopper species have camouflage
colouring and patterning, although some have
bright "warning" coloration and produce noxious
chemicals. The antennae are always short. The
females, which are nearly always larger than the
males, do not have a conspicuous ovipositor.
• **LIFE-CYCLE** Males sing during
the day to attract females. After mating,
egg masses are laid in the ground. The
eggs are protected by a foamy substance
that is secreted by the female.
• **OCCURRENCE** Worldwide,
especially in warm areas.
On the ground and
among vegetation.
• **REMARK** Many
grasshoppers are agricultural pests. The
most notorious pest species – locusts –
can form huge swarms that cause
widespread devastation.

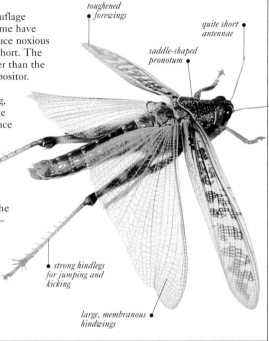

toughened
forewings

quite short
antennae

saddle-shaped
pronotum

strong hindlegs
for jumping and
kicking

large, membranous
hindwings

SCHISTOCERCA GREGARIA, the
African Desert Locust, is one of the
world's most destructive insects. Vast
swarms can devour up to 100,000
tonnes of food in a day.

Length 1–8cm (⅜–3¼in), most 1.5–3cm (⅝–1¼in)	Feeding habits

Order ORTHOPTERA	Family EUMASTACIDAE	No. of species 1,200

MONKEY-HOPPERS

The head of these slim insects
is long and set at an angle to the
thorax. Many are brightly coloured,
while some resemble leaves or
sticks. The hindlegs are thin and
elongated, with distinctive spines
on the lower half of the tibiae. At
rest, many species sit with their
hindlegs splayed out sideways.
• **LIFE-CYCLE** Mating occurs with
the male on top of the female. Eggs
are laid in the ground or in detritus.
• **OCCURRENCE** S.E. Asia,
Africa, India, North and South
America, especially in tropical and
subtropical regions. In a variety of
habitats, including woodland, forests,
and grassland.

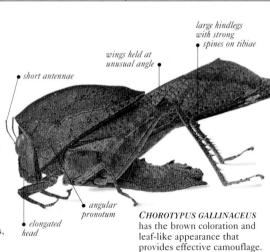

large hindlegs
with strong
spines on tibiae

wings held at
unusual angle

short antennae

angular
pronotum

elongated
head

CHOROTYPUS GALLINACEUS
has the brown coloration and
leaf-like appearance that
provides effective camouflage.

Length 1.5–5.8cm (⅝–2¼in)	Feeding habits

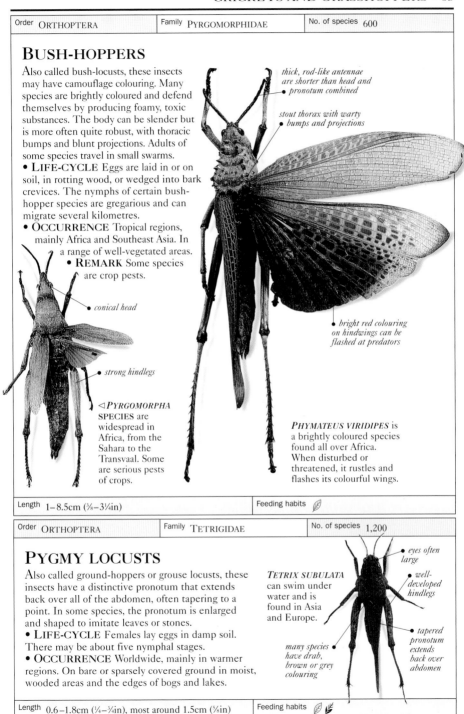

Order ORTHOPTERA	Family PYRGOMORPHIDAE	No. of species 600

BUSH-HOPPERS

Also called bush-locusts, these insects may have camouflage colouring. Many species are brightly coloured and defend themselves by producing foamy, toxic substances. The body can be slender but is more often quite robust, with thoracic bumps and blunt projections. Adults of some species travel in small swarms.
• LIFE-CYCLE Eggs are laid in or on soil, in rotting wood, or wedged into bark crevices. The nymphs of certain bush-hopper species are gregarious and can migrate several kilometres.
• OCCURRENCE Tropical regions, mainly Africa and Southeast Asia. In a range of well-vegetated areas.
• REMARK Some species are crop pests.

thick, rod-like antennae are shorter than head and pronotum combined

stout thorax with warty bumps and projections

conical head

strong hindlegs

bright red colouring on hindwings can be flashed at predators

◁ *PYRGOMORPHA* SPECIES are widespread in Africa, from the Sahara to the Transvaal. Some are serious pests of crops.

PHYMATEUS VIRIDIPES is a brightly coloured species found all over Africa. When disturbed or threatened, it rustles and flashes its colourful wings.

Length 1–8.5cm (⅜–3¼in)	Feeding habits

Order ORTHOPTERA	Family TETRIGIDAE	No. of species 1,200

PYGMY LOCUSTS

Also called ground-hoppers or grouse locusts, these insects have a distinctive pronotum that extends back over all of the abdomen, often tapering to a point. In some species, the pronotum is enlarged and shaped to imitate leaves or stones.
• LIFE-CYCLE Females lay eggs in damp soil. There may be about five nymphal stages.
• OCCURRENCE Worldwide, mainly in warmer regions. On bare or sparsely covered ground in moist, wooded areas and the edges of bogs and lakes.

TETRIX SUBULATA can swim under water and is found in Asia and Europe.

eyes often large

well-developed hindlegs

many species have drab, brown or grey colouring

tapered pronotum extends back over abdomen

Length 0.6–1.8cm (¼–¾in), most around 1.5cm (⅝in)	Feeding habits

STICK AND LEAF INSECTS

T HERE ARE 3 FAMILIES and 2,500 species of stick and leaf insects in the order Phasmatodea. The largest of the three families is the Phasmatidae.

These long, slow-moving, herbivorous insects are mostly nocturnal. By day, they protect themselves from predators with their highly convincing stick- or leaf-like appearance. If disturbed, many stick insects remain motionless, holding their legs tightly along their body. Males tend to be smaller than females. The males of many species are winged, whereas females are often wingless. Most females drop, scatter, or flick their eggs from the end of the abdomen. A few lay eggs in soil or glue them to plants. Some eggs are very seed-like and attract ants, who take them back to the safety of their nests. Metamorphosis is incomplete.

Order PHASMATODEA	Family PHASMATIDAE	No. of species 2,450

STICK INSECTS

Also called walking sticks, these night-feeding insects are usually brown or green and often spiny or warty. Females are frequently wingless and males are often winged. The wings may be short, or short, tough forewings may protect much larger, membranous, fan-shaped hindwings.
• **LIFE-CYCLE** Eggs are deposited from the abdomen. They are laid in soil or stuck to plants. The tiny first-stage nymphs usually rely on camouflage coloration for survival.
• **OCCURRENCE** Mainly in tropical areas and some warm, temperate regions. Among vegetation or on the foliage of shrubs and trees.
• **REMARK** Defence tactics include using noises, smells, postures, and coloration, or shedding legs if seized by predators (the legs often grow back later).

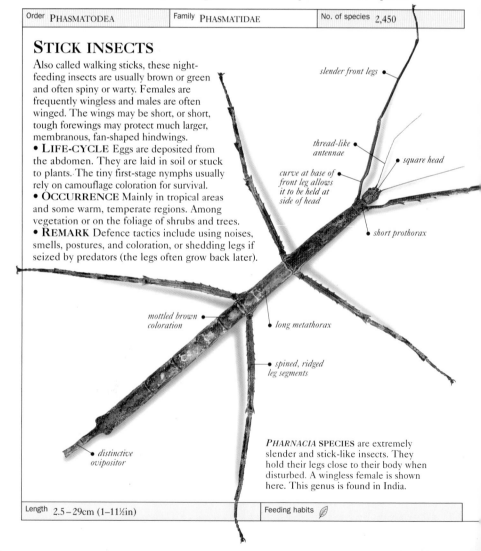

slender front legs •

thread-like • antennae

curve at base of • front leg allows it to be held at side of head

• square head

• short prothorax

mottled brown • coloration

• long metathorax

• spined, ridged leg segments

• distinctive ovipositor

PHARNACIA SPECIES are extremely slender and stick-like insects. They hold their legs close to their body when disturbed. A wingless female is shown here. This genus is found in India.

Length 2.5–29cm (1–11½in)	Feeding habits 🌿

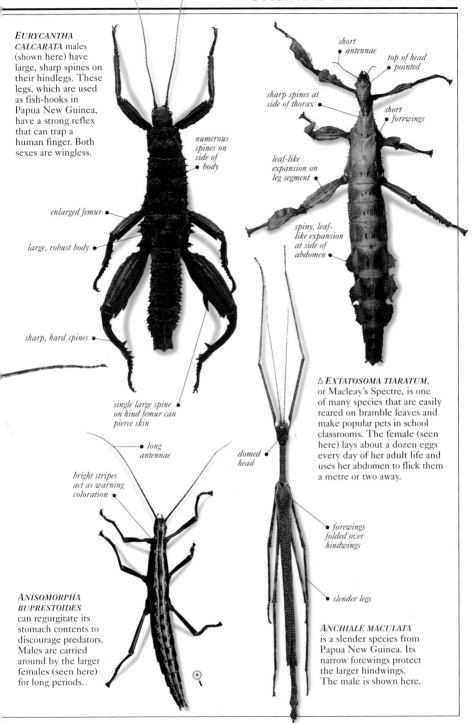

EURYCANTHA CALCARATA males (shown here) have large, sharp spines on their hindlegs. These legs, which are used as fish-hooks in Papua New Guinea, have a strong reflex that can trap a human finger. Both sexes are wingless.

numerous spines on side of body

enlarged femur

large, robust body

sharp, hard spines

single large spine on hind femur can pierce skin

long antennae

short antennae

top of head pointed

sharp spines at side of thorax

short forewings

leaf-like expansion on leg segment

spiny, leaf-like expansion at side of abdomen

domed head

△ *EXTATOSOMA TIARATUM*, or Macleay's Spectre, is one of many species that are easily reared on bramble leaves and make popular pets in school classrooms. The female (seen here) lays about a dozen eggs every day of her adult life and uses her abdomen to flick them a metre or two away.

bright stripes act as warning coloration

forewings folded over hindwings

slender legs

ANISOMORPHA BUPRESTOIDES can regurgitate its stomach contents to discourage predators. Males are carried around by the larger females (seen here) for long periods.

ANCHIALE MACULATA is a slender species from Papua New Guinea. Its narrow forewings protect the larger hindwings. The male is shown here.

Order PHASMATODEA	Family PHYLLIIDAE	No. of species 30

LEAF INSECTS

Flat, expanded abdomens, extended leg segments, and brown or green coloration all give these insects a resemblance to living or dead leaves. When they are at rest, their veined forewings may cover their transparent hindwings to complete the disguise, made even more effective by surface texturing, blotches of colour, and an ability to sway in the breeze. The antennae are short and smooth in females, longer and slightly hairy in males.
• LIFE-CYCLE The seed-like eggs are dropped on to the ground, where they hatch. The young nymphs feed nocturnally.
• OCCURRENCE Seychelles, Southeast Asia, Northern Queensland, and New Guinea. In any well-vegetated habitat.

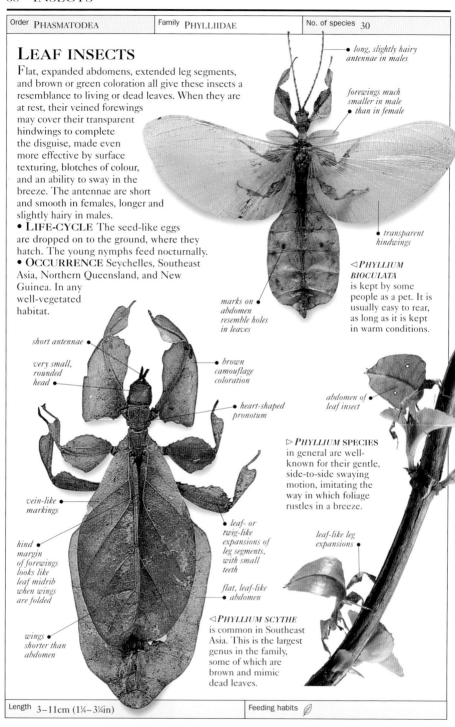

long, slightly hairy antennae in males

forewings much smaller in male than in female

transparent hindwings

◁ PHYLLIUM BIOCULATA is kept by some people as a pet. It is usually easy to rear, as long as it is kept in warm conditions.

marks on abdomen resemble holes in leaves

short antennae

very small, rounded head

brown camouflage coloration

heart-shaped pronotum

abdomen of leaf insect

▷ PHYLLIUM SPECIES in general are well-known for their gentle, side-to-side swaying motion, imitating the way in which foliage rustles in a breeze.

vein-like markings

hind margin of forewings looks like leaf midrib when wings are folded

leaf- or twig-like expansions of leg segments, with small teeth

flat, leaf-like abdomen

leaf-like leg expansions

wings shorter than abdomen

◁ PHYLLIUM SCYTHE is common in Southeast Asia. This is the largest genus in the family, some of which are brown and mimic dead leaves.

Length 3–11cm (1¼–3¼in)	Feeding habits

EARWIGS

THE ORDER DERMAPTERA is relatively small. It is divided into 10 families, containing about 1,900 species. Commonly known as earwigs, these relatively flat insects have short, veinless forewings that protect the large, fan-shaped hindwings. The abdomen is mobile and telescopic, with a pair of forceps-like appendages that are usually straight in females and curved in males.

Metamorphosis is incomplete. Females typically lay their eggs in soil, although some parasitic species give birth to live nymphs. The females show a high degree of maternal care, for example licking fungal spores off the eggs and guarding them from predators. This care continues for some time after the eggs hatch. Females feed their nymphs by bringing food into the nest or by regurgitating part of their own meals. Eventually, the nymphs have to disperse – as they grow, the mother starts regarding them as a potential meal.

Earwigs moult up to five times. Apart from increasing in size and gaining antennal segments with each moult, they look similar to their parents.

Earwigs like confined spaces. Their name may refer to the popular belief that they enter human ears (they rarely do), or to the shape of their hindwings.

Order DERMAPTERA	Family CARCINOPHORIDAE	No. of species 400

CARCINOPHORID EARWIGS

Most of these short-legged species lack hind-wings, and in some species the short, toughened forewings are also absent. Many species are dark-coloured in shades of dark brown, black, or reddish with paler yellow or red markings. The antennae have fewer than 20 segments, and the abdominal forceps, which are typically short, may not be symmetrical in males.
• LIFE-CYCLE Eggs are laid in burrows or in leaf-litter.
• OCCURRENCE Worldwide, especially in warmer regions. In a wide range of habitats.
• REMARK Some species of earwigs are significant pests of cultivated flowers and fruits.

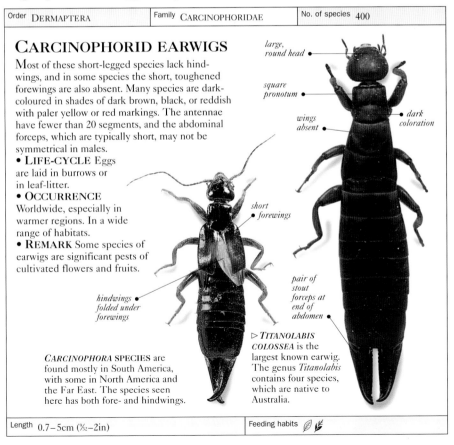

large, round head

square pronotum

wings absent

dark coloration

short forewings

hindwings folded under forewings

pair of stout forceps at end of abdomen

CARCINOPHORA SPECIES are found mostly in South America, with some in North America and the Far East. The species seen here has both fore- and hindwings.

▷ *TITANOLABIS COLOSSEA* is the largest known earwig. The genus *Titanolabis* contains four species, which are native to Australia.

Length 0.7–5cm (⅜–2in)	Feeding habits

Order DERMAPTERA	Family FORFICULIDAE	No. of species 450

COMMON EARWIGS

These typically slender earwigs vary in appearance but are usually dark brown or blackish brown, with paler legs and thread-like antennae. The abdominal forceps of the male earwigs are highly curved, whereas those of the females are relatively straight.

• LIFE-CYCLE Females usually lay eggs in soil under rocks or bark. They guard the eggs against predators and lick them clean to prevent any fungal growth. Apart from plant matter, the diet may include small caterpillars, aphids, and other insects.

• OCCURRENCE Worldwide. In leaf-litter and soil, under bark, or in crevices.

• REMARK These insects can be pests of crops or garden plants.

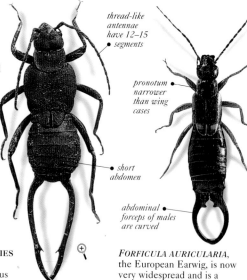

thread-like antennae have 12–15 segments

pronotum narrower than wing cases

short abdomen

abdominal forceps of males are curved

ALLODAHLIA SPECIES are found in India, Malaysia, and various parts of the Far East.

FORFICULA AURICULARIA, the European Earwig, is now very widespread and is a well-known pest of flowers.

Length 1.2–2.5cm (½–1in)	Feeding habits

Order DERMAPTERA	Family LABIDURIDAE	No. of species 75

STRIPED EARWIGS

Also called long-horned earwigs, due to their long antennae, these relatively robust insects are reddish brown. They are usually winged, although some species are wingless. Common species have dark stripes on the pronotum and wing cases.

• LIFE-CYCLE Female striped earwigs lay their eggs in deep tunnels that they have dug in sandy soil.

• OCCURRENCE Worldwide, mainly in warmer regions. Around coastal areas, riverbanks, and mudflats; in leaf-litter, sand, and debris, or under stones.

• REMARK Some species can discharge a foul-smelling liquid produced by abdominal glands.

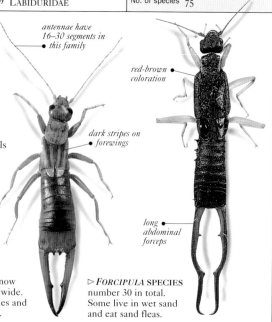

antennae have 16–30 segments in this family

red-brown coloration

dark stripes on forewings

long abdominal forceps

LABIDURA RIPARIA is now found practically worldwide. It lives on sandy beaches and along shores and rivers.

▷ *FORCIPULA* SPECIES number 30 in total. Some live in wet sand and eat sand fleas.

Length 1.2–4cm (½–1⅝in)	Feeding habits

MANTIDS

T HE ORDER MANTODEA includes 8 families and 2,000 species. The common name "praying mantis" is derived from the distinctive way in which the front legs are held up and together, as if in prayer.

Variable in shape, mantids have several clearly recognizable features: a triangular head with large, forward-facing eyes, and an elongated prothorax with the front pair of legs distinctively modified for catching live prey. Mantids are excellent hunters. The head is very mobile – these are the only insects that can turn their head to "look over their shoulder". The forward-pointing eyes give true binocular vision and allow distances to be calculated accurately, while the front legs can be used to pounce and seize prey very quickly – typically in less than a tenth of a second.

Mantids are mostly daytime fliers, eating a wide range of arthropods and even vertebrates such as frogs and lizards. Many avoid attacks by bats with the aid of a special ultrasonic ear on the underside of the thorax.

Metamorphosis is incomplete. The eggs are laid inside a papery or foam-like case, which is fixed to twigs or other surfaces. The females of some species guard this egg case from predators.

Order MANTODEA	Family EMPUSIDAE	No. of species 30

EMPUSIDS

These insects are often large and slender in shape. Most empusid species have highly distinctive leaf- or lobe-like expansions on the ends of the femora and at the sides of the abdomen. The front tibiae are distinctive in having long spines that alternate with two to four short ones. The head has a lobe-like outgrowth, and the antennae appear plumed in males.

• LIFE-CYCLE Typical of mantids, the eggs are laid inside an egg case, and the first nymphs moult very soon after emerging.

• OCCURRENCE Africa, parts of the Mediterranean region, and Asia. On flowers, foliage, and other vegetation.

• REMARK As with all mantids, the size and activity of prey are carefully judged. Prey capable of fighting back are not attacked if they are more than half the size of the mantid.

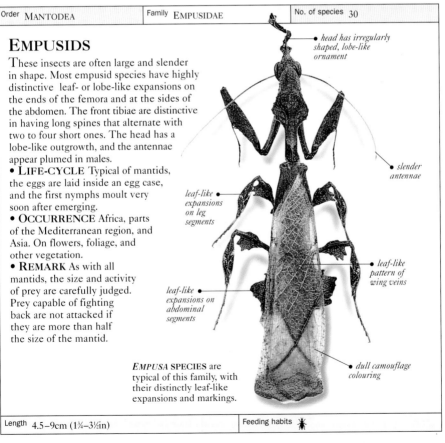

• head has irregularly shaped, lobe-like ornament

• slender antennae

leaf-like • expansions on leg segments

leaf-like • expansions on abdominal segments

• leaf-like pattern of wing veins

• dull camouflage colouring

EMPUSA SPECIES are typical of this family, with their distinctly leaf-like expansions and markings.

Length 4.5–9cm (1¾–3½in)	Feeding habits 🐜

Order MANTODEA	Family HYMENOPODIDAE	No. of species 50

FLOWER MANTIDS

Vivid coloration, including bright pinks and greens, coupled with body ornamentation, allows these species to blend in perfectly with the flowers on which they rest, awaiting the arrival of prey. Some parts of the body, such as the legs, often have broad extensions that resemble leaves. The forewings may have coloured bands or spirals, or conspicuous, circular marks that look rather like eye-spots. In female flower mantids, the wings are sometimes short.

• LIFE-CYCLE Eggs are laid in a case, typically attached to vegetation. Young nymphs tackle small prey as soon as their cuticle has hardened.

• OCCURRENCE In tropical regions all around the world, except Australia. On a wide variety of vegetation.

pointed ornamentation on head

broad, leaf-like extensions on legs

◁ *CREOBROTER* SPECIES are typical of the flower mantids in appearance. They are particularly abundant in India and across Southeast Asia generally.

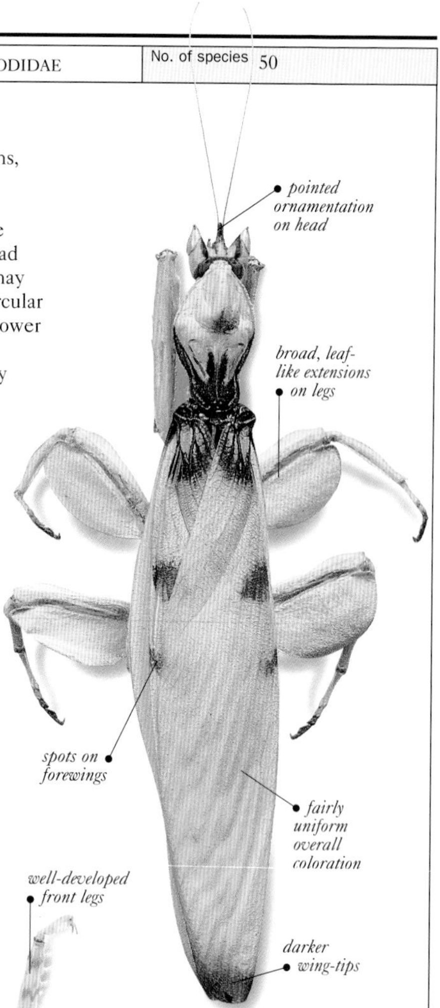

coloration blends in with flowers

spots on forewings

distinctive eye-spot markings on wings

well-developed front legs

fairly uniform overall coloration

darker wing-tips

large, forward-pointing eyes

△ *HYMENOPUS CORONATUS* usually has flower-mimicking camouflage colouring. The coloration varies enormously, depending on the immediate surroundings in which a nymph develops – red, for example, if a nymph spends its early life on a red bloom.

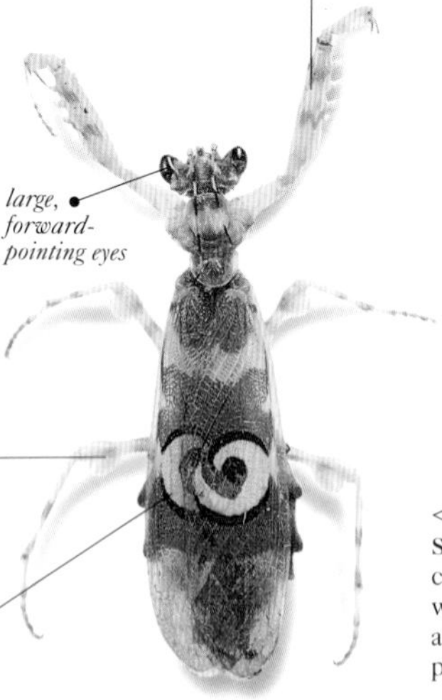

leaf-like expansions on legs

◁ *PSEUDOCREBOTRA* SPECIES have vividly coloured markings on their wings. These can be flashed as a warning when any potential predators threaten.

bright eye-spots with black margins

Length 2–8cm (¾–3¼in)	Feeding habits 🐜

Order MANTODEA	Family MANTIDAE	No. of species 1,400

COMMON PRAYING MANTIDS

Many members of this family vary very little in appearance, and most have green or brown coloration. Unlike families such as the flower mantids (see opposite), the wings of common praying mantids are seldom patterned, although they may have wing-spots.
• LIFE-CYCLE The female praying mantid is famed for eating the male after mating. However, although mating can be dangerous for the smaller males, they are cautious and are seldom eaten by their partners in the wild. Anything from a few dozen to a couple of hundred eggs are laid. The eggs are surrounded by a papery or foam-like egg case.
• OCCURRENCE Worldwide, mainly in warmer regions. On any kind of vegetation where there is a good supply of prey.

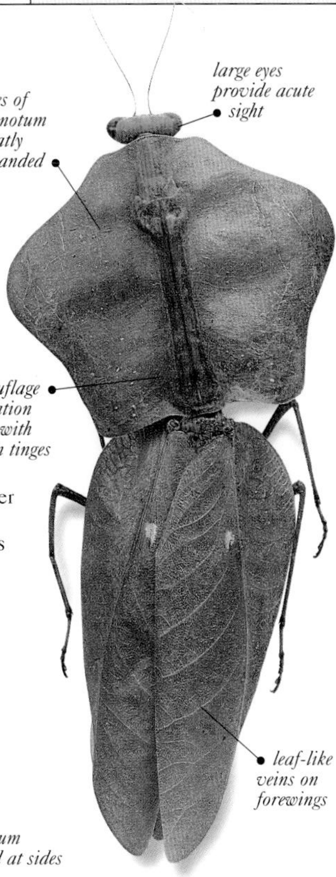

large eyes provide acute sight

sides of pronotum greatly expanded

camouflage coloration green with brown tinges

▷ *CHOERADODIS STALII* and other large species in this genus are capable of preying on vertebrates such as salamanders and frogs.

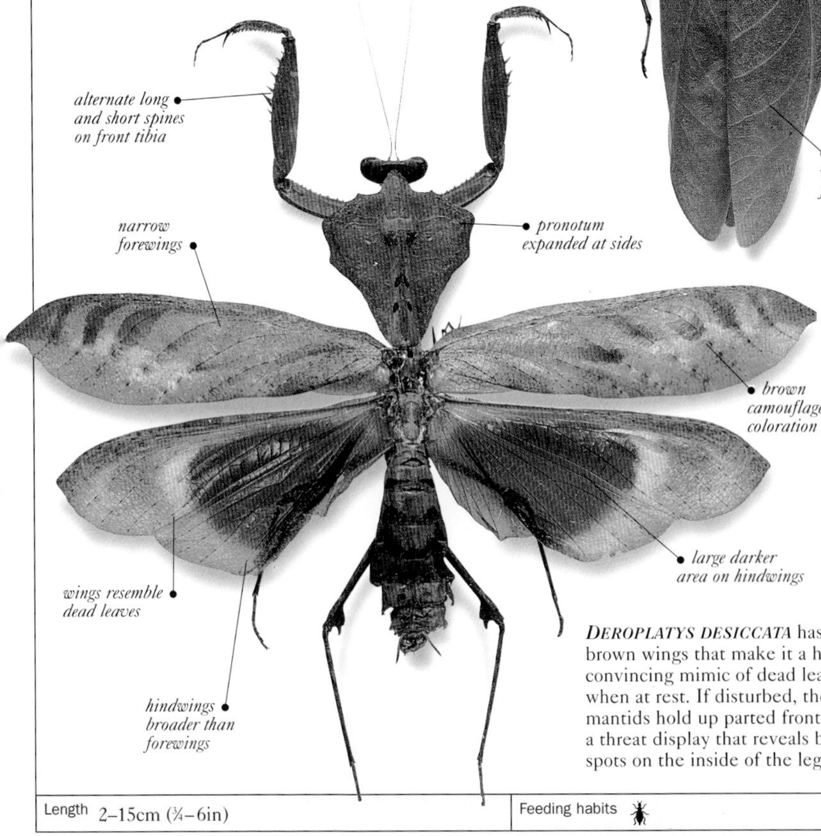

leaf-like veins on forewings

alternate long and short spines on front tibia

narrow forewings

pronotum expanded at sides

brown camouflage coloration

large darker area on hindwings

wings resemble dead leaves

hindwings broader than forewings

DEROPLATYS DESICCATA has brown wings that make it a highly convincing mimic of dead leaves when at rest. If disturbed, these mantids hold up parted front legs in a threat display that reveals bright spots on the inside of the legs.

Length 2–15cm (¾–6in)	Feeding habits 🐜

COCKROACHES

T HE ORDER BLATTODEA contains 6 families and 4,000 species. These leathery insects usually have an oval, flat shape that lets them squeeze through tight spaces. The head is often covered by a shield-like pronotum. There are generally two pairs of wings. The forewings are usually tough and cover large, membranous hindwings.

Most cockroaches live in tropical regions. Less than one per cent of them are pests, adapted to human habitats, thriving in warm, unsanitary conditions, and often carrying disease. Cockroaches mostly eat dead or decaying organic matter, including bird and bat guano.

Their main defence against predators is that they are highly sensitive to vibration and can run fast, but some also spray or ooze toxic chemicals.

Female cockroaches produce sexual pheromones to which males respond, and males may produce aphrodisiac secretions to encourage the female. Mating occurs back to back, and sperm is transferred in a packet. The females lay up to 40 eggs, which are surrounded by a tough case that may be dropped, stuck to the ground, or carried about, partly projecting from the end of the cockroach's abdomen. Metamorphosis in these insects is incomplete.

Order BLATTODEA	Family BLABERIDAE	No. of species 1,000

LIVE-BEARING COCKROACHES

These large insects often have well-developed, pale brown wings with dark markings. In many species of large live-bearing cockroach, however, the females are wingless and burrow under wood and stones.

• **LIFE-CYCLE** In some species, courtship rituals involve the production of sounds. Males of *Nauphoeta cinerea*, for example, make sounds by rubbing a ridged part of the forewing against the hind corners of the pronotum. Most species reproduce by producing eggs that develop inside an egg case. The egg case is fully extended from the end of the female's abdomen, rotated by 90 degrees, and then drawn back into the body to be brooded.

• **OCCURRENCE** Tropical regions. In a range of habitats including rainforests and caves.

• **REMARK** Some species are reared as laboratory animals.

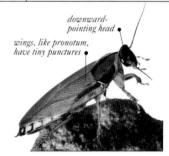

downward-pointing head

wings, like pronotum, have tiny punctures

PYCNOSCELIS SURINAMENSIS is a widespread tropical pest. It lives underground and is often found in caves, in bat droppings.

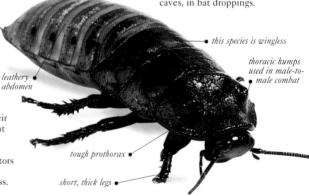

this species is wingless

thoracic humps used in male-to-male combat

leathery abdomen

GROMPHADORHINA PORTENTOSA males use their large thoracic humps to fight each other. Also known as the Madagascan Hissing Cockroach, it startles predators by squeezing air out of its spiracles to create a loud hiss.

tough prothorax

short, thick legs

Length 2.5–6cm (1–2½in), rarely up to 9.5cm (3¾in)	Feeding habits

Order BLATTODEA	Family BLATTIDAE	No. of species 600

COMMON COCKROACHES

Most species in this family are brown, red-brown, or black-brown, with varied markings. They are highly active, can run very fast, and fly in hot conditions. They usually hide during the day, coming out to feed after dark. Some species produce repellent chemicals that can cause skin rashes and temporary blindness. They eat a range of organic matter and food scraps, regurgitating already digested food and leaving behind their faeces and a characteristic unpleasant smell.
• **LIFE-CYCLE** Males may have special glands on their back, which secrete chemicals that lure a mate into a suitable position for copulation. Females can produce up to 50 egg cases, each containing 12–14 eggs. The egg cases may be deposited on, buried in, or stuck to a variety of surfaces.
• **OCCURRENCE** Mostly in tropical and subtropical regions. Pest species are found around ports, warehouses, sewers, and rubbish dumps.

head pointed downwards

slender legs

◁ *NEOSTYLOPYGA RHOMBIFOLIA*, the Harlequin Cockroach, produces a chemical (amyl acetate) that smells like pear drops, which it uses to defend itself against predators.

distinctive black-and-cream patterns

protruding egg case

△ *BLATTA ORIENTALIS*, known as the Oriental Cockroach, is a common domestic pest. Like many species, it has spread beyond its native region by travelling on board ships.

tough, leathery forewings

long, sensitive antennae

cerci

protruding egg case held until suitable hiding place for it is found

spiny legs

long hindlegs

PERIPLANETA AMERICANA, the American Cockroach, is often used in laboratories because it is easy to culture. Originally from Africa, it is now found worldwide.

Length 2 – 4.5cm (¾–1¾in)	Feeding habits

Order BLATTODEA	Family BLATTELLIDAE	No. of species 1,750

BLATTELLIDS

Many of these insects are pale brown but a few are olive-green. Darker brown markings are common, typically on the wings and pronotum. They often appear shiny. Both sexes usually have fully developed wings.

• **LIFE-CYCLE** Adults and nymphs are scavengers. Females can be very fertile, breeding five to six times a year and producing about 40 eggs each time. In many species, eggs are carried in a case protruding from the abdomen, which is dropped to the ground just before they hatch.

• **OCCURRENCE** Worldwide, especially in warmer areas. In woodland litter, debris, waste tips, and buildings.

dark marks on pronotum

long, thread-like antennae

brownish coloration

△ **BLATTELLA GERMANICA**, the German Cockroach, is a common pest of houses, kitchens, and stores. Females can produce thousands of eggs in a lifetime.

shiny body surface

no wings

pale side margins

△ **LOBOPTERA DECIPIENS** is a wingless species, found in Europe. It lives in leaf-litter and under stones and bark.

males and females fully winged

MEGALOBLATTA LONGIPENNIS, from Peru, Ecuador, and Panama, is the largest winged cockroach in the world. Its wingspan can measure up to 20cm (8in).

Length 0.8–10cm (⁵⁄₁₆–4in), most under 2cm (¾in)	Feeding habits

Order BLATTODEA	Family CRYPTOCERCIDAE	No. of species 4

WOOD ROACHES

The wood roaches are the most primitive cockroaches and belong to a single genus: *Cryptocercus*. They are quite elongate and are shiny dark brown to black.

• **LIFE-CYCLE** Wood roaches live in small family groups comprising two adults and up to 20 nymphs. They tunnel through dead wood – sticking their eggs to the roof of the wooden chambers – and use special symbiotic protozoans in their gut to break down the wood cellulose into sugars. Each time nymphs moult, they eat some of the group's droppings to make sure they are reinfected with the protozoans.

• **OCCURRENCE** North America and China. In dead wood in a variety of habitats.

head tucked under pronotum

CRYPTOCERCUS PUNCTULATUS, the American Wood Roach, probably encompasses several separate species.

short, spiny legs

Length 2–3cm (¾–1¼in)	Feeding habits

WEB-SPINNERS

A RELATIVELY SMALL GROUP, the order Embioptera contains just 8 families and 300 species. Web-spinners are gregarious insects, whose common name is derived from their ability to make expansive silk tunnels in soil, in litter, and under bark. These are used for protection against predators.

Web-spinners' legs are short. Adults and young of all species have a swollen tarsal segment on their front legs, which contains silk glands. The silk is ejected through the many bristle-like structures found on the underside of the segments. As the insect moves its front feet against a surface, a silk sheet gradually forms.

Females are wingless, while the males typically have two pairs of narrow wings. The females and nymphs remain within the colony, but adult males fly off to find mates in other colonies. Metamorphosis in this order is incomplete.

Order EMBIOPTERA	Family CLOTHODIDAE	No. of species 14

CLOTHODIDS

The members of this family are typically long and cylindrical in shape, with short legs. They have simple biting mouthparts. The front legs of both adult and young clothodids have swollen tarsal segments that contain many silk glands. The eyes are small, and the antennae have 10–35 segments.

• LIFE-CYCLE Males do not feed but use their mandibles to hold on to females during mating. There is a great deal of maternal care. The female covers her eggs with silk and bits of detritus. After the eggs hatch, the female may feed her young pre-chewed food.

• OCCURRENCE Worldwide, in tropical and subtropical regions. In a wide variety of habitats, from rainforest to desert.

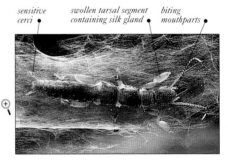

sensitive cerci • *swollen tarsal segment containing silk gland* • *biting mouthparts* •

CLOTHODA URICHI, seen here inside its silken web, is a native of Trinidad and is either communal or solitary. Both types live in the same silk nest as their nymphs, but communal females produce more eggs than solitary females.

Length 0.5–2cm (³⁄₁₆–³⁄₄in)	Feeding habits

Order EMBIOPTERA	Family EMBIIDAE	No. of species 250

EMBIIDS

These colony-living web-spinners are usually quite stout. Males have two pairs of wings that bend forwards, as is typical of the order. This allows them to move backwards in the colony's galleries. When flying, a substance is pumped into a special wing vein to stiffen each wing.

• LIFE-CYCLE Eggs are laid in silk-lined galleries. The female guards the eggs and young until they disperse.

• OCCURRENCE South America and Africa. In various habitats, in bark, tree holes, and cracks, and under stones and rocks.

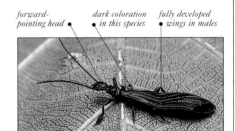

forward-pointing head • *dark coloration in this species* • *fully developed wings in males* •

EMBIID SPECIES have very similar characteristics to members of the family Clothodidae (see above).

Length 0.5–2.2cm (³⁄₁₆–³⁄₄in)	Feeding habits

TERMITES

T HE 7 FAMILIES AND 2,750 species in the order Isoptera are never found more than 50 degrees north or south of the Equator. A typical termite is pale, soft, and wingless, with chewing or biting mandibles and short antennae, but the different castes within a colony have varying features. The reproductives, including kings and queens, have two pairs of long wings, short cerci, and round or oval heads. The sterile soldiers have proportionately larger heads than the workers. Eyes are often reduced or absent. Metamorphosis is incomplete.

Termites live in mud nests or vast underground mazes with huge ventilation chimneys above ground. Most eat dead or rotting wood – this is the only order where all families digest cellulose – and many attack crops or wooden building timbers.

Order ISOPTERA	Family HODOTERMITIDAE	No. of species 19

HARVESTER TERMITES

These termites have eyes, and their mandibles have large, distinct teeth on the inner surfaces. They are cream or light to dark brown in colour. The pronotum is saddle-shaped, extending down at the sides.
• LIFE-CYCLE Eggs are laid in the colony's nest, which may be up to 6m (20ft) under the ground. Workers forage above ground for grass or small bits of wood to feed the young.
• OCCURRENCE Africa and Asia. In regions of dry savannah. In soil.
• REMARK These termites can be pests of open pasture. They may eat the food of larger herbivores – both wild animals and domestic cattle – and encourage soil erosion.

strong, toothed mandibles •   *large, tough head* •   *three thoracic segments of similar size* •

HODOTERMES SPECIES are widespread in African savannahs and always make underground nests. During the day, they can be seen running along the ground, collecting grass and pieces of twigs.

Length 0.4–1cm (⅛₂–⅜in) – soldiers and workers only	Feeding habits

Order ISOPTERA	Family RHINOTERMITIDAE	No. of species 345

SUBTERRANEAN TERMITES

The pronotum of all castes is rounded at the back and may, in some species, appear almost heart-shaped. Soldiers have no eyes. Coloration is cream or light to dark brown.
• LIFE-CYCLE Eggs are laid in the colonies, which are found either in soil or in damp wood that touches the ground.
• OCCURRENCE Worldwide, in warm regions. In various habitats, in soil and wood.
• REMARK Some species are timber pests. One North American species is such a serious pest that there are special building regulations to help prevent the damage it can cause.

soft, pale body •   *small thoracic segments* •   *rounded or bluntly heart-shaped pronotum* •   *bead-like antennal segments* •

RETICULITERMES LUCIFUGUS makes its nest under the ground, inside damp wood. These pale, soft-bodied termites, with distinctive antennae, are found throughout southern Europe.

Length 2–8mm (¹⁄₁₆–⁵⁄₁₆in) – soldiers and workers only	Feeding habits

Order ISOPTERA	Family TERMITIDAE	No. of species 1,950

HIGHER TERMITES

This highly variable family of pale cream to dark brown termites comprises almost three-quarters of all termites. Workers and soldiers have no eyes, and soldiers often have large, biting mandibles or a snout-like head from which sticky poisons are ejected.

• LIFE-CYCLE Eggs are laid in nests that vary from small structures in trees and soil mounds to vast underground mazes. Some queens are enormous and may produce several thousand eggs a day.

• OCCURRENCE Worldwide, in tropical and subtropical areas. In varied habitats, in trees or soil.

• REMARK Many higher termites are pests. This family has a more complex, rigid caste system than other, "lower" termites.

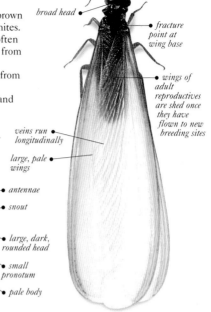

broad head •

• fracture point at wing base

• wings of adult reproductives are shed once they have flown to new breeding sites

veins run • longitudinally

large, pale • wings

• antennae

• snout

• large, dark, rounded head

• small pronotum

• pale body

TRINERVITERMES GEMINATUS, or the Snouted Harvester Termite, has soldiers that can produce a repellent secretion from a gland in their head. This is used against their main enemies – ants.

MACROTERMES SPECIES can attack cocoa and coconut crops. At certain times of year, reproductive adults like this one emerge from colonies in their thousands.

Length 0.4–1.4cm (⁵⁄₃₂–⁹⁄₁₆in) – soldiers and workers only	Feeding habits 

Order ISOPTERA	Family TERMOPSIDAE	No. of species 20

TRUE DAMP WOOD TERMITES

These termites, pale to dark brown in colour, are also called rotten wood termites. The flat pronotum is much narrower than the head.

• LIFE-CYCLE Eggs are laid in nests in decaying wood – usually wood that is in contact with the ground.

• OCCURRENCE Warm regions of North and South America, Africa, Asia, and Australia. Mainly in rotting trees or fallen logs.

• REMARK A few species are pests of structural timbers, especially those buried in the ground, such as telegraph poles.

large wings shed after mating and establishing new colony •

rounded head •

small, flat • pronotum

newly emerged • reproductive termites are pale and white as they have not yet hardened

ZOOTERMOPSIS ANGUSTICOLLIS, the Pacific Damp Wood Termite, is native to North America, where it can become a troublesome pest of damp timber found above the ground.

Length 3–6mm (⅛–¼in) – soldiers and workers only	Feeding habits

ANGEL INSECTS

T HE ORDER ZORAPTERA consists of a single family divided into 29 species. Discovered in the early twentieth century, these small, delicate, termite-like insects are light straw to dark brown or blackish in colour. The adults of most species come in two forms. One form has no eyes or ocelli, and is pale and wingless like the nymphs. The other form has eyes and three ocelli, dark coloration, and two pairs of wings with minimal venation. The winged forms are responsible for dispersal to new locations when the habitat becomes unsuitable for some reason. When a new site is found, the wings are shed. Both wingless and winged angel insects have short abdominal cerci and unspecialized, downward-pointing mouthparts, which resemble those seen in crickets and grasshoppers (see pp.60–65). The tarsi are divided into two segments.

Metamorphosis is incomplete, and the nymphs vary in appearance depending on whether or not they have wings as adults. Native to North and South America, Africa, and parts of eastern Asia, these insects live gregariously under the bark of rotting trees and in wood dust and damp leaf-litter. Some are associated with termite colonies.

Order ZORAPTERA	Family ZOROTYPIDAE	No. of species 29

ANGEL INSECTS

These insects are pale and wingless, or darker and winged. The body has an oval abdomen, and the distinctly triangular head carries a pair of antennae with nine bead-like segments.
• **LIFE-CYCLE** Both nymphs and adults eat fungal threads and small arthropods, especially mites. There may be quite complex sexual behaviour, and the males of some species give the females mating gifts in the form of secretions from glands in their heads. In other species, the males fight for mates, kicking each other with their hindlegs. Success in mating seems to increase with age and not simply with size. Eggs are laid where these insects live, in places such as rotting wood or leaf-litter.
• **OCCURRENCE** Worldwide in tropical and warm temperate regions, except Australia. In rotting wood, sawdust piles, and leaf-litter, and under tree bark.
• **REMARK** All the known species are contained in the single genus, *Zorotypus*.

antennae have nine segments

triangular head

short, fairly stout legs

short, one-segmented cerci

ZOROTYPUS HUBBARDI is pale to dark brown in real life – this is a slide-mounted specimen that has been stained red. Native to eastern and southern USA, this species is found in sawdust and rotting logs and under dead bark.

Length 2–3mm (1⁄16–1⁄8in)	Feeding habits

BARKLICE AND BOOKLICE

COMMONLY KNOWN as barklice and booklice, the order Psocoptera contains 35 families and 3,000 species. Drab and soft-bodied, these common insects are often overlooked. The large head has a bulbous forehead, bulging eyes, and long, thread-like antennae. The thorax appears humped when seen from the side, and there are usually two pairs of membranous wings, held roof-like over the body when folded. A few species bear live young, but most lay eggs. Metamorphosis is incomplete, and there are usually five nymphal stages.

These insects are found in a range of terrestrial habitats, from soil to tree canopies, where they eat algae, lichens, moulds, and fungal spores. Some eat pollen and plant tissues. Many species are solitary but some can be gregarious.

Order PSOCOPTERA	Family CAECILIIDAE	No. of species 330

CAECILIIDS

Most caeciliids are brown, yellow, or green in colour and either fully winged or short-winged – very few are wingless. The wings may have markings, and the front edge of the forewings and wing veins are covered in short hairs.
• LIFE-CYCLE Typically, a batch of 12–16 eggs is laid on leaves. Where there is more than one generation each year, the autumn-laid eggs fall with the leaves on to the leaf-litter and hatch out in spring.
• OCCURRENCE Worldwide. In a range of habitats. Commonly in the foliage of deciduous trees and in low-growing vegetation and grasses. A few are found among leaf-litter.

veins of forewings edged with red-brown •   *yellow body* •   *brown central stripe on head* •

CAECILIUS FLAVIDUS is abundant on the leaves of many deciduous trees. This species is native to Europe, including the United Kingdom.

Length 1.5–4mm (1/32–5/32in)	Feeding habits

Order PSOCOPTERA	Family ECTOPSOCIDAE	No. of species 150

ECTOPSOCIDS

These pale brown species may have dark brown bands on their abdomen. The head is covered with hairs. They have fairly small eyes and 13-segmented antennae that are quite long in males but shorter in females.
• LIFE-CYCLE Females lay batches of eggs on the veins of dead or withering leaves and cover them with silk threads produced by special glands. In good conditions – an hospitable, disease-free environment with ample food – populations can become very large. The adult-like nymphs are hairy and feed on fungal threads, rotting matter, algae, and sometimes on pollen.
• OCCURRENCE Worldwide. In various habitats. Among the dry leaves of various deciduous tree species or in leaf-litter. Some species occur in greenhouses or in houses.

antennae have 13 segments •   *dark brown body* •

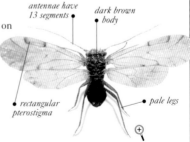

rectangular pterostigma •   • *pale legs*

ECTOPSOCUS BRIGGSI is a European species, found on deciduous trees. Most members of this species have fully developed wings, although short-winged forms occur occasionally.

Length 1.5–3.5mm (1/32–5/32in)	Feeding habits

Order PSOCOPTERA	Family LIPOSCELIDAE	No. of species 150

LIPOSCELID BOOKLICE

These insects have flat bodies and distinctively swollen femora on their hindlegs. Most liposcelid booklice have no wings, although winged forms do occur. The head has a pair of short antennae and small eyes.
• LIFE-CYCLE The female lice lay their eggs in places such as leaf-litter, crevices in tree bark, and birds' nests. Their nymphs look much like small adults although they have shorter antennae.
• OCCURRENCE Worldwide. In dry leaf-litter, under bark, and in nests. Some are found inside buildings and food stores.
• REMARK Some species can be pests. Several species in the genus *Liposcelis* will thrive in damp conditions and can attack stored flour, cereals, books, and papers. They have also been known to do considerable damage to museum collections of plants and insects.

slender, multi-segmented antennae • *small eyes* • *short legs* • *swollen hind femora* •

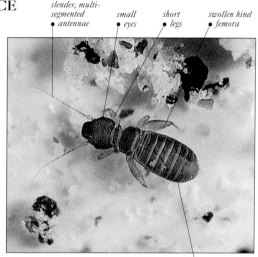

LIPOSCELIS TERRICOLIS is a very widely distributed species of booklice. It is found in leaf-litter and in a variety of stored, dry produce, where it can become a pest.

• *flat, yellowish brown body*

Length 0.5–1.5mm (¹⁄₄₄–¹⁄₃₂in)	Feeding habits

Order PSOCOPTERA	Family PSOCIDAE	No. of species 500

COMMON BARKLICE

Most common barklice have dull brown, grey, or blackish coloration, and many species have pale markings. The wings are hairless and may also be mottled with spots and irregular patches.
• LIFE-CYCLE The females lay their eggs singly or in groups, usually inside crevices in tree bark. The life-cycle of a typical barklouse may take a little over two months; there are usually several generations in a year. Most species feed on fungi, pollen, algae, and lichen.
• OCCURRENCE Worldwide. On the bark, branches, and twigs of a wide variety of trees and shrubs. Some of the commonest species in this order belong to this family, and enormous populations can be found on certain trees.

wings held, roof-like, over body • *brown mottling* • *shiny, humped thorax* •

PSOCOCERASTIS GIBBOSA is native to Europe and parts of Asia. It is found on a number of tree species and is the largest member of its genus in the United Kingdom.

• *downward-pointing mandibles*

Length 1–6mm (¹⁄₃₂–¹⁄₄in)	Feeding habits

PARASITIC LICE

THE 25 FAMILIES and 6,000 species in the order Phthiraptera are commonly called parasitic lice. These flattened-looking insects are wingless ectoparasites, living permanently on the bodies of birds and mammals without killing them. The mouthparts are used for biting skin, feather, or fur but in the sucking lice they are used exclusively for feeding on blood. The legs are short and are often modified for clinging on to either fur or feathers. Metamorphosis is incomplete in this order.

Different lice are linked with specific hosts, such as the species with distinctive mouthparts that are found only on warthogs and African elephants. Many lice are also restricted to certain areas of the body, and so more than one species can inhabit a host at the same time.

Order PHTHIRAPTERA	Family MENOPONIDAE	No. of species 650

BIRD LICE

These lice have oval abdomens and short, stout legs. The large, roughly triangular head has biting mandibles.
• LIFE-CYCLE Eggs are glued singly to feathers. The majority of species feed on feather fragments, but some also take blood and skin secretions.
• OCCURRENCE Worldwide. On a variety of birds.
• REMARK Some species, such as *Menopon gallinae* (the Shaft Louse), can be serious poultry pests.

broad head •

MENACANTHUS STRAMINEUS, the Chicken Body Louse, is a widespread species. Infestation can lead to feather-loss and infection.

• *two claws on each leg*

Length 1–6mm (½₂–¼in), most under 4mm (⅗₂in)	Feeding habits 🌢 ✱

Order PHTHIRAPTERA	Family PEDICULIDAE	No. of species 2

HUMAN LICE

These lice are small, pale, and elongate, with short, strongly clawed legs for gripping on to their hosts. The small head bears distinctively dark eyes. The human louse, *Pediculus humanus*, also occurs on some monkeys. It has two subspecies: *P. humanus* subsp. *corporis* (the Body Louse) and *P. humanus* subsp. *capitis* (the Head Louse). The other species in the family, *P. schaeffi*, is found exclusively on apes.
• LIFE-CYCLE The Body Louse lives and lays eggs in the fibres of clothing, whereas the Head Louse lives entirely in hair and glues its eggs (nits) to hair.
• OCCURRENCE Worldwide. On humans, apes, and monkeys.
• REMARK Outbreaks of head lice are common among young children. Resistance to insecticidal shampoos is developing and regular washing and fine combing is often just as effective. Up until World War II, many more soldiers died of louse-borne epidemic typhus and relapsing fever than were ever killed in battle.

narrow head •

strong, curved legs

PEDICULUS HUMANUS CAPITIS, like all human lice, has mouthparts specially adapted for sucking blood.

pear-shaped, flattened body

body gorged with blood •

blood meal visible through body wall •

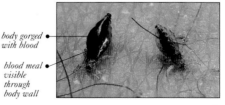

PEDICULUS HUMANUS CORPORIS, the Body Louse, lives in people's clothing and can transmit the organism that causes typhus.

Length 1.5–3.5mm (½₂–⅛in)	Feeding habits 🌢

Order PHTHIRAPTERA	Family PTHIRIDAE	No. of species 2

PUBIC LICE

These pale to translucent lice have a squat, flat body and a head that is very much narrower than the thorax. The middle and hindlegs are especially stout and have strong claws for gripping on to hair shafts. The family consists of the Human Pubic Louse, *Pthirus pubis*, and the Gorilla Pubic Louse, *Pthirus gorillae*.

• **LIFE-CYCLE** After mating, the female uses a strong, waterproof glue to stick her eggs singly to pubic hairs. Both nymphs and adults feed on the host's blood, leaving bluish marks on the skin.

• **OCCURRENCE** Wherever their hosts live. The Human Louse is found worldwide.

• **REMARK** Contrary to popular belief, these very slow-moving lice do not jump and can transfer to new hosts only during intimate contact.

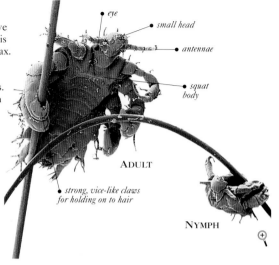

eye

small head

antennae

squat body

ADULT

strong, vice-like claws for holding on to hair

NYMPH

PTHIRUS PUBIS, the Human Pubic Louse, may be found in armpits and beards as well as the groin. Although unpleasant, it is not known to transmit disease.

Length 1.5–2.5mm (¹⁄₁₆–³⁄₃₂in)	Feeding habits 🌢

Order PHTHIRAPTERA	Family TRICHODECTIDAE	No. of species 350

MAMMAL CHEWING LICE

Generally pale brown in colour, these lice have large, square heads with conspicuous, short antennae and distinctive mandibles. The legs are short, each with a single tarsal claw. Females have blunt-ended abdomens, while those of males are slightly pointed. The name comes from the fact that these lice live on mammals such as horses, cattle, sheep, goats, dogs, and cats, as well as non-domesticated species.

• **LIFE-CYCLE** Eggs are laid on the hairs or fleece of hosts. This is also where the nymphs live, eating tiny pieces of skin, hair, sebaceous gland secretions, and, sometimes, blood.

• **OCCURRENCE** Worldwide. On their mammalian hosts.

• **REMARK** The Dog Louse carries tapeworms. Infestations affect animals' health as they spend more time grooming than feeding.

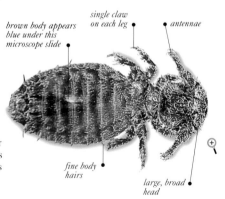

single claw on each leg

antennae

brown body appears blue under this microscope slide

fine body hairs

large, broad head

DAMALINIA ORIS is a small louse found on sheep all over the world. A badly infected sheep may carry over 750,000 lice.

Length 1–3mm (¹⁄₃₂–¹⁄₈in)	Feeding habits 🌢

BUGS

THE ORDER HEMIPTERA consists of 134 families and 82,000 species. Bugs range from minute, wingless insects to giant water bugs that can catch fish and frogs. They are found in all terrestrial habitats, in fresh water, and even on the surface of southern oceans.

Modern classification recognizes four suborders: the Coleorrhyncha (just one family); the Heteroptera (true bugs); the Auchenorrhyncha (plant-, leaf-, and tree-hoppers, lantern bugs, and cicadas); and the Sternorrhyncha (jumping plant lice, whiteflies, aphids, and scale insects). The selection below is divided into true bugs (Acanthosomatidae to Tingidae);

hoppers and relatives (Aphrophoridae to Membracidae); and aphids and relatives (Aleyrodidae to Psyllidae).

Only the true bugs include predacious and blood-sucking species; the others are herbivorous. All bugs have piercing, sucking mouthparts in the form of a long rostrum made up of slender stylets sheathed by the labium. Many bugs that suck plant sap are serious crop pests.

Metamorphosis is incomplete. Eggs are laid in and on vegetation and soil, and under bark. There are typically about five nymphal stages. Some bugs produce live young, and others reproduce without the need for males.

Order HEMIPTERA	Family ACANTHOSOMATIDAE	No. of species 250

ACANTHOSOMATID BUGS

These bugs are similar to their close relations, the shield-backed bugs (see p.94). They are usually green, brown, or grey, although some are brightly coloured. The head is small, and the body is oval or tapers behind the wide pronotum.
• LIFE-CYCLE Eggs are laid on host shrubs and trees. Nymphs and adults suck sap from these plants.
• OCCURRENCE Worldwide, especially in warm regions. In various habitats such as woods, meadows, and scrubland, wherever their host plants are found.
• REMARK Some females show maternal care, guarding the eggs until they hatch. Small nymphs may shelter under their mother's body when predators threaten.

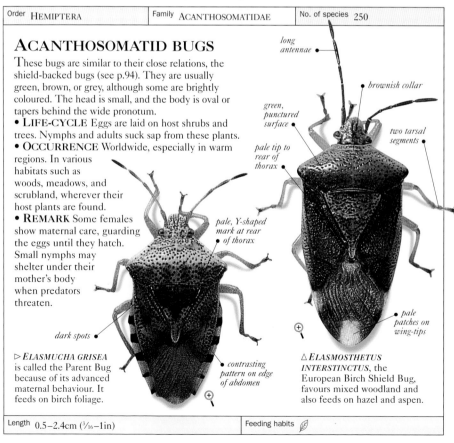

long antennae

brownish collar

green, punctured surface

two tarsal segments

pale tip to rear of thorax

pale, Y-shaped mark at rear of thorax

dark spots

pale patches on wing-tips

▷ *ELASMUCHA GRISEA* is called the Parent Bug because of its advanced maternal behaviour. It feeds on birch foliage.

contrasting pattern on edge of abdomen

△ *ELASMOSTHETUS INTERSTINCTUS*, the European Birch Shield Bug, favours mixed woodland and also feeds on hazel and aspen.

Length 0.5–2.4cm (³⁄₁₆–1in)	Feeding habits

Order HEMIPTERA	Family ALYDIDAE	No. of species 300

BROAD-HEADED BUGS

These slender, drably coloured bugs look similar to squash bugs (see p.88) but their heads are broader and almost as long as the pronotum. Many species are ant-like and are found in association with ants, although other species resemble wasps. Broad-headed bugs have well-developed scent glands, which can produce even stronger odours than those emitted by stink bugs (see p.92).

• LIFE-CYCLE Eggs are laid in soil and leaf-litter. Some nymphs have been found in ants' nests.

• OCCURRENCE Worldwide. On grasses and leguminous plants.

• REMARK Some species are pests of rice and millet.

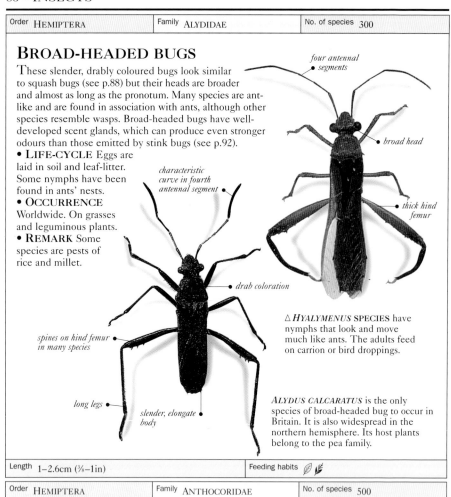

four antennal segments

broad head

characteristic curve in fourth antennal segment

thick hind femur

drab coloration

spines on hind femur in many species

long legs

slender, elongate body

△ HYALYMENUS SPECIES have nymphs that look and move much like ants. The adults feed on carrion or bird droppings.

ALYDUS CALCARATUS is the only species of broad-headed bug to occur in Britain. It is also widespread in the northern hemisphere. Its host plants belong to the pea family.

Length 1–2.6cm (⅜–1in)	Feeding habits

Order HEMIPTERA	Family ANTHOCORIDAE	No. of species 500

FLOWER BUGS

Also known as minute pirate bugs, species in this family have relatively flat bodies that can be elongate or oval. Most flower bugs have fully developed wings, but the wings are short in some species. The antennae have four segments.

• LIFE-CYCLE Eggs are laid in small groups on the undersides of leaves or leaf stalks. They may also be laid in stems, bark, or debris.

• OCCURRENCE Worldwide. Under bark and in flowers, vegetation, leaf-litter, and fungi. Some live in animal nests and burrows.

• REMARK These bugs are used to control aphids, thrips, scale insects, moths, and mites.

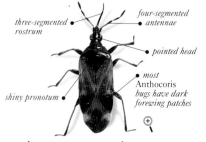

three-segmented rostrum

four-segmented antennae

pointed head

shiny pronotum

most Anthocoris bugs have dark forewing patches

ANTHOCORIS SPECIES lay eggs on leaves infested with prey such as aphids. Anthocoris bugs are attracted to odours given off by their prey.

Length 2–5mm (¹⁄₁₆–³⁄₁₆in)	Feeding habits

Order HEMIPTERA	Family ARADIDAE	No. of species 1,800

BARK BUGS

Also known as flat bugs, these insects have a flat, oval shape and many are wingless. They always have black or dark to reddish brown camouflage colouring, and the upper surface of the body is very rough or has short projections or dimples.
• LIFE-CYCLE Eggs are laid in a small mass under bark or in leaf-litter. A few species care for their young, and the male may share this role.
• OCCURRENCE Worldwide. Most under bark, especially of decaying trees; among plant debris and leaf-litter on the forest floor; among certain fungi.

▷*ARADUS BETULAE* is a European species, typically found underneath the bark of birch trees.

short, stout antennae

long, piercing mouthparts are coiled inside head when not in use

flat, oval body

thorax covered with fine dimples and coarse ridges

◁*ARADUS ATERRIMUS* adults and nymphs are often found together under the bark of a variety of deciduous trees.

Length 0.3–1.3cm (⅛–½in)	Feeding habits

Order HEMIPTERA	Family BELOSTOMATIDAE	No. of species 150

GIANT WATER BUGS

Also called electric light bugs due to their attraction to lights after dark, this family contains the largest bugs. They are oval, flat, and streamlined with enlarged front legs that are modified as prey-capturing pincers. The flat middle and hindlegs are usually fringed with hairs to aid swimming. The antennae are small and concealed.
• LIFE-CYCLE Males can attract mates by using their bodies to send waves across the water's surface. Females may glue the eggs to the male's back or to vegetation. In back-brooding species, male giant water bugs care for the eggs and keep them moist.
• OCCURRENCE Worldwide, especially in subtropical and tropical regions. In slow-moving streams, pools, and ponds.
• REMARK Nymphs and adults are highly predacious, even catching frogs, fish, and small birds.

LETHOCERUS GRANDIS is one of the giant water bug species that are caught and eaten by humans in certain parts of Southeast Asia.

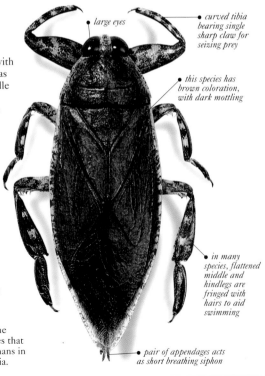

large eyes

curved tibia bearing single sharp claw for seizing prey

this species has brown coloration, with dark mottling

in many species, flattened middle and hindlegs are fringed with hairs to aid swimming

pair of appendages acts as short breathing siphon

Length 1.5–10cm (⅝–4in)	Feeding habits

| Order HEMIPTERA | Family CIMICIDAE | No. of species 90 |

BED BUGS

These flat, oval, wingless bugs are usually reddish or brown. As well as humans, many species use other mammals and birds as hosts.
• LIFE-CYCLE Eggs are laid in crevices in the adults' resting place. Each of the five nymphal stages needs a huge meal of blood. The life-cycle spans two to ten months.
• OCCURRENCE Worldwide. On hosts, in nests and caves, and in crevices in buildings.

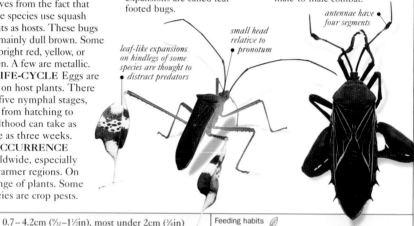

flattened, oval shape • • non-functional, scale-like wing remnants

CIMEX LECTULARIUS is the best-known species, long associated with humans. It feeds at night and finds hosts partly by sensing their body heat.

| Length 3–6mm (⅛–¼in) | Feeding habits 🔵 |

| Order HEMIPTERA | Family COREIDAE | No. of species 2,000 |

SQUASH BUGS

The name of this family derives from the fact that some species use squash plants as hosts. These bugs are mainly dull brown. Some are bright red, yellow, or green. A few are metallic.
• LIFE-CYCLE Eggs are laid on host plants. There are five nymphal stages, and from hatching to adulthood can take as little as three weeks.
• OCCURRENCE Worldwide, especially in warmer regions. On a range of plants. Some species are crop pests.

DIACTOR SPECIES and related genera with leaf-like expansions are called leaf-footed bugs.

leaf-like expansions on hindlegs of some species are thought to • distract predators

small head relative to • pronotum

THASUS ACUTANGULUS has very strong hindlegs, used in male-to-male combat.

antennae have • four segments

| Length 0.7–4.2cm (%₃₂–1½in), most under 2cm (¾in) | Feeding habits |

| Order HEMIPTERA | Family CORIXIDAE | No. of species 550 |

WATER BOATMEN

These swimming bugs rest under the water surface or cling to plants. Usually dark red- or yellow-brown, their short front legs form a food scoop; the slim middle legs grip plants; and the oar-like hind pair are used for swimming.
• LIFE-CYCLE Boatmen attract mates by rubbing body parts together. Eggs are glued to submerged objects and plants.
• OCCURRENCE Worldwide. In pools, ponds, lakes, and slow streams.

▽ CALLICORIXA WOLLASTONI is probably predacious and in some areas is found in peaty upland pools.

large, dark • eyes

hindlegs adapted for swimming •

△ CORIXA PUNCTATA is adapted for swimming, like its relatives, but will also fly readily.

| Length 0.3–1.5cm (⅛–⅝in) | Feeding habits |

Order HEMIPTERA	Family GELASTOCORIDAE	No. of species 90

TOAD BUGS

All members of this family can jump on to their prey, and many have a bumpy or warty appearance. Oval and broad in shape, with bulging, toad-like eyes, their camouflage colouring blends in with mud, sand, or shingle.
• LIFE-CYCLE Eggs are laid in plant debris or wet soil. The nymphs have good sight.
• OCCURRENCE Worldwide, especially in the southern hemisphere. By ponds and streams; in rotten wood and leaf-litter; and under stones.

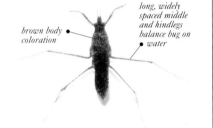

bulging eyes in broad head

front legs used to seize prey

camouflage colouring

NERTHRA GRANDICOLLIS is a typical toad bug in appearance. This common genus is the largest in the family.

slender hindlegs

Length 0.7–1.4cm (⁹⁄₃₂–⁵⁄₈in)	Feeding habits 🐜

Order HEMIPTERA	Family GERRIDAE	No. of species 500

POND-SKATERS

Also known as water-striders, these very fast-moving, often wingless bugs are adapted to living on the surface of water. The body is dark brown or black and thickly covered with velvety, water-repelling hairs. Long middle and hindlegs spread the bug's weight evenly over the water surface.
• LIFE-CYCLE "Ripple communication" is used to attract mates. Egg masses are laid on submerged plants or floating objects or inserted into plant stems.
• OCCURRENCE Worldwide. In water-bodies ranging from small pools and ponds to streams, rivers, lakes, and warm oceans.

long, widely spaced middle and hindlegs balance bug on water

brown body coloration

GERRIS SPECIES are widespread and may be found living on running water. Attracted by the ripples made by prey, they move rapidly to seize their quarry on the surface.

Length 0.2–3.5cm (¹⁄₁₆–1¼in), most 1–1.5cm (³⁄₈–⁵⁄₈in)	Feeding habits 🐜

Order HEMIPTERA	Family HYDROMETRIDAE	No. of species 120

WATER-MEASURERS

Also called marsh-treaders, these delicate bugs are elongate, slender, with thread-like legs. They are reddish to dark brown in colour. The head is very long, with protruding eyes about halfway along its length. Most water-measurers are wingless, but winged forms sometimes occur in certain species.
• LIFE-CYCLE The long eggs are laid singly, glued to vegetation or pond edges, above the water.
• OCCURRENCE Worldwide, especially in tropical and subtropical regions. On marginal or floating plants of ponds, pools, marshes, and swamps.
• REMARK Nymphs and adults prefer prey that is injured or freshly dead. They are particularly fond of mosquito eggs and similar immobile food items.

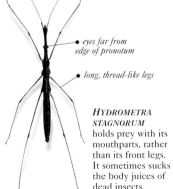

eyes far from edge of pronotum

long, thread-like legs

HYDROMETRA STAGNORUM holds prey with its mouthparts, rather than its front legs. It sometimes sucks the body juices of dead insects.

Length 0.3–2.2cm (⅛–¾in)	Feeding habits 🐜

Order HEMIPTERA	Family MIRIDAE	No. of species 8,000

PLANT BUGS

These elongate or oval bugs are usually quite delicate and fragile. The different species display a great diversity of colours and markings.

• **LIFE-CYCLE** Eggs are typically inserted inside plant tissues. There are five nymphal stages. The nymphs – and adults – of most species suck plant sap, but many will attack small, soft-bodied prey such as aphids and scale insects.

• **OCCURRENCE** Worldwide. In a wide range of well-vegetated habitats.

• **REMARK** Many plant bug species are pests, damaging crops such as grasses, cotton, coffee, and potatoes. Some predacious species have been used as biological control agents of red spider mites on fruit trees.

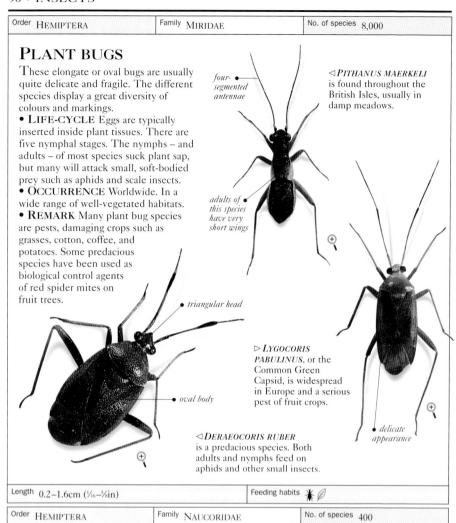

four-segmented antennae

adults of this species have very short wings

◁ *PITHANUS MAERKELI* is found throughout the British Isles, usually in damp meadows.

triangular head

oval body

▷ *LYGOCORIS PABULINUS*, or the Common Green Capsid, is widespread in Europe and a serious pest of fruit crops.

◁ *DERAEOCORIS RUBER* is a predacious species. Both adults and nymphs feed on aphids and other small insects.

delicate appearance

Length 0.2–1.6cm (¹⁄₁₆–⅝in)	Feeding habits

Order HEMIPTERA	Family NAUCORIDAE	No. of species 400

SAUCER BUGS

Also called creeping water bugs, these flat insects have a smooth oval or rounded body. Most are dark brown to black. The front legs are used to catch prey, while the hindlegs are used for swimming.

• **LIFE-CYCLE** Eggs are inserted, in rows, into the stems of aquatic plants.

• **OCCURRENCE** Worldwide. In ponds, streams, and, occasionally, hot springs.

• **REMARK** Air from the surface is trapped under the folded wings, helping the bugs to stay alive and buoyant as they move about feeding under the water.

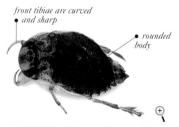

front tibiae are curved and sharp

rounded body

ILYOCORIS CIMICOIDES is found in Europe. Despite having fully developed wings, this species rarely, if ever, flies.

Length 0.6–1.8cm (¼–¾in)	Feeding habits

Order HEMIPTERA	Family NEPIDAE	No. of species 250

WATER SCORPIONS

These greyish brown or reddish brown bugs are split into two main genera. *Nepa* species are oval and flattened with fairly short legs while *Ranatra* species are elongate with relatively long legs. The strong front legs are modified for catching prey while the other legs are used for walking. There is a long breathing siphon at the end of the abdomen.

• **LIFE-CYCLE** Females often lay eggs inside plant tissues. Adults and nymphs may communicate by rubbing the base of their legs against their body to make sounds.

• **OCCURRENCE** Worldwide, especially in warmer regions. In slow-moving streams, ponds, pools, and bogs.

• **REMARK** Most species have wings but rarely fly.

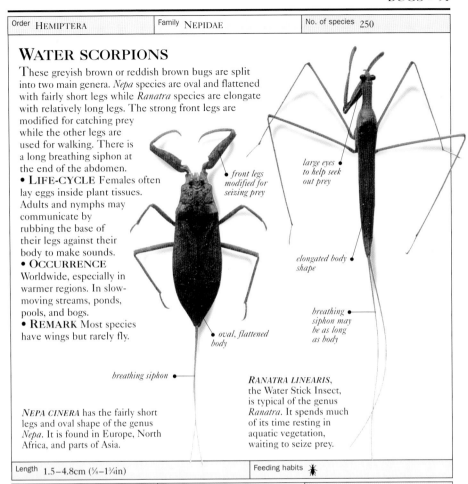

front legs modified for seizing prey

large eyes to help seek out prey

elongated body shape

breathing siphon may be as long as body

oval, flattened body

breathing siphon

NEPA CINERA has the fairly short legs and oval shape of the genus *Nepa*. It is found in Europe, North Africa, and parts of Asia.

RANATRA LINEARIS, the Water Stick Insect, is typical of the genus *Ranatra*. It spends much of its time resting in aquatic vegetation, waiting to seize prey.

Length 1.5–4.8cm (⅗–1⅞in)	Feeding habits 🦗

Order HEMIPTERA	Family NOTONECTIDAE	No. of species 350

BACK-SWIMMERS

The adults of these compact bugs are good fliers. Their upper body surface is typically pale and convex, with a central ridge. The underside, which faces up as they swim, is normally dark brown or black. The front and middle legs are used to catch prey, and the long, hair-fringed hindlegs are used for swimming.

• **LIFE-CYCLE** Males make mate-attracting noises by rubbing part of the rostrum against the front legs. Eggs are laid in batches of less than ten at a time, inserted into aquatic plants.

• **OCCURRENCE** Worldwide. In small pools and ponds, and at the edges of lakes.

large, dark, and shiny eyes *rostrum* *long hindlegs used for swimming*

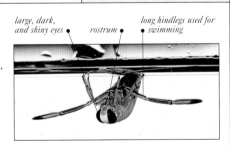

NOTONECTA GLAUCA, the Common Back-swimmer, is widespread in the ponds and ditches of the United Kingdom. It is sensitive to vibration and uses its large eyes to locate approaching prey.

Length 0.2–1.7cm (¹⁄₁₆–⅝in)	Feeding habits 🦗

| Order HEMIPTERA | Family PENTATOMIDAE | No. of species 5,500 |

STINK BUGS

Also known as shield bugs because of their distinctive shape, these insects can produce strong defensive odours from their thoracic glands. Many species are green or brown but some are brightly coloured.
• LIFE-CYCLE Females lay clusters of barrel-shaped eggs on plants. There are five nymphal stages. Many nymphs start life as herbivores but later become predators or mixed feeders.
• OCCURRENCE Worldwide. On vegetation, shrubs, and trees.

front of head rounded

square pronotum with sharp point at each side

sides of abdomen banded black and orange

reddish legs

PENTATOMA RUFIPES, the Forest Bug, feeds on deciduous trees, such as oak, in European woodland.

black dorsal surface of head

black legs

shiny body surface

red pronotum with black patches

large, black, semicircular mark on hind part of thorax

three pale spots on front edge of hind part of thorax

uniform pale green colour

slightly sculptured body surface

EURYDEMA DOMINULUS has bold coloration that warns predators of its distastefulness. It is found throughout Europe.

NEZARA VIRIDULA, also called the Green Vegetable Bug or Green Stink Bug, is a worldwide pest of fruit, vegetables, and cereals.

| Length 0.5–2.5cm (³⁄₁₆–1in) | Feeding habits |

| Order HEMIPTERA | Family PYRRHOCORIDAE | No. of species 400 |

RED BUGS

Also called fire bugs due to their bright black, red, and orange colouring, these conspicuous insects often have quite elongate bodies.
• LIFE-CYCLE Eggs are laid in damp soil or debris, or among the seeds of the host plant. Many species feed communally on the seeds and juices of their host plants.
• OCCURRENCE Worldwide, especially in warmer regions. On plants belonging to the family Malvaceae and others.
• REMARK Some species in the genus *Dysdercus*, called cotton-stainers, are pests of cotton crops. The staining of the cotton boll is caused by a fungus introduced when the bugs feed.

long antennae

rostrum usually held under body

elongate body

LOHITA GRANDIS, the Giant Red Bug, is a pest in forested parts of northern India. It feeds on various trees, including two species that are valuable sources of timber.

bold coloration

| Length 0.8–2.2cm (⁵⁄₁₆–⁷⁄₈in); some up to 5cm (2in) | Feeding habits |

Order HEMIPTERA	Family REDUVIIDAE	No. of species 6,000

ASSASSIN BUGS

These bugs get their name from being highly predacious. They vary from being stout-bodied to very elongate with thread-like legs. Most species are dark-coloured but some may have bright markings. The head has a short, curved, three-segmented rostrum. The front legs are strong and shorter than the others – ideal for gripping prey.

• **LIFE-CYCLE** Up to 50 eggs are laid in cracks or crevices or in soil, or are glued to foliage. They may be guarded by the males.

• **OCCURRENCE** Worldwide, especially in subtropical and tropical regions. In a wide variety of habitats.

• **REMARK** Certain blood-sucking species carry Chagas' disease, which can cause heart failure.

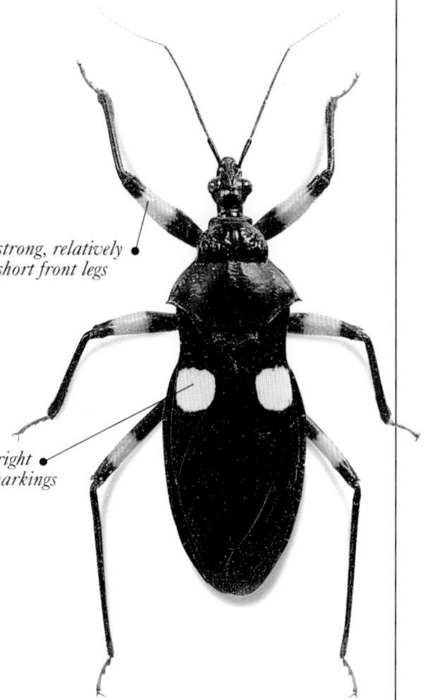

strong, relatively short front legs

bright markings

long, slender legs

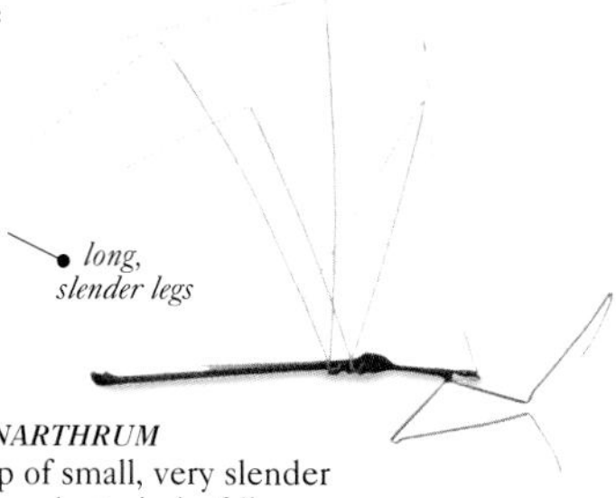

GARDENA MELANARTHRUM
belongs to a group of small, very slender species that live on plants, in leaf-litter, and in caves. Some live in spider webs, feeding on the trapped prey.

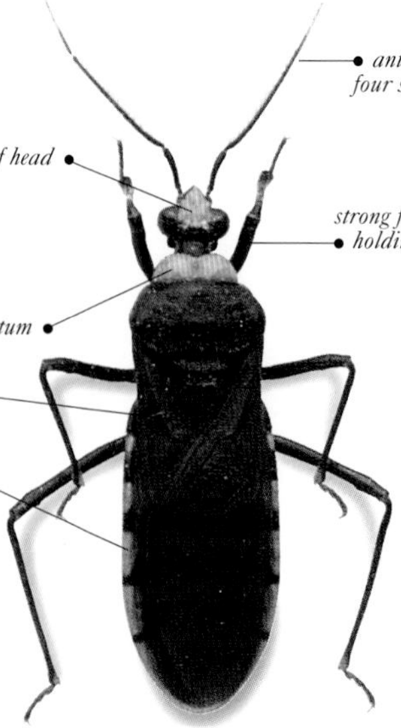

PLATYMERIS BIGUTTATA is a large species that is kept in laboratories and as a pet. Care should be taken as its toxic saliva can cause temporary blindness.

antennae have four segments

yellow upper surface of head

strong front legs for holding prey

short, curved, three-segmented rostrum

yellow pronotum

dark body

black and orange bands around edge of abdomen

***CENTRASPIS* SPECIES** are native to Mozambique and Guinea. The three species in this genus hunt for insect prey on the ground and among vegetation.

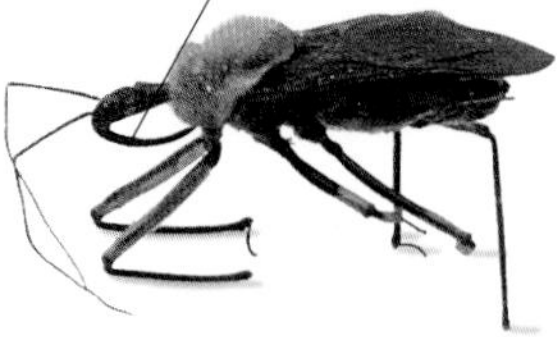

RHINOCORIS ALLUAUDI, like many assassin bugs, can make sounds by rasping its rostrum on a special file on the underside of the thorax.

Length 0.7–4cm (⁹⁄₃₂–1½in)	Feeding habits 🐜 💧

Order HEMIPTERA	Family SALDIDAE	No. of species 300

SHORE BUGS

These small, oval bugs are mostly brown or black in colour. The head often has a long rostrum and large eyes with a notch in the hind margin. Shore bugs can jump to escape predators.
• LIFE-CYCLE Females lay eggs at the base of various grasses or in moss. Some nymphs burrow.
• OCCURRENCE Worldwide. On muddy margins of salt marshes, streams, and ponds. On the seashore, among vegetation and seaweed.

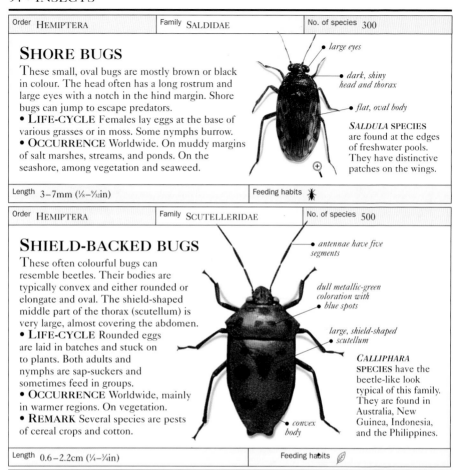

large eyes

dark, shiny head and thorax

flat, oval body

SALDULA SPECIES are found at the edges of freshwater pools. They have distinctive patches on the wings.

Length 3–7mm (⅛–⁹⁄₃₂in)	Feeding habits

Order HEMIPTERA	Family SCUTELLERIDAE	No. of species 500

SHIELD-BACKED BUGS

These often colourful bugs can resemble beetles. Their bodies are typically convex and either rounded or elongate and oval. The shield-shaped middle part of the thorax (scutellum) is very large, almost covering the abdomen.
• LIFE-CYCLE Rounded eggs are laid in batches and stuck on to plants. Both adults and nymphs are sap-suckers and sometimes feed in groups.
• OCCURRENCE Worldwide, mainly in warmer regions. On vegetation.
• REMARK Several species are pests of cereal crops and cotton.

antennae have five segments

dull metallic-green coloration with blue spots

large, shield-shaped scutellum

CALLIPHARA SPECIES have the beetle-like look typical of this family. They are found in Australia, New Guinea, Indonesia, and the Philippines.

convex body

Length 0.6–2.2cm (¼–⅞in)	Feeding habits

Order HEMIPTERA	Family TINGIDAE	No. of species 2,000

LACE BUGS

These small, greyish bugs are distinguished by lace-like patterning and sculpturing on the upper surface of the wings. The pronotum can extend, hood-like, over the head.
• LIFE-CYCLE Eggs are inserted into the tissue of host plants. The females of some species show complex maternal care of both their eggs and young.
• OCCURRENCE Worldwide. On herbaceous plants and trees.
• REMARK Many species are pests, but some are used as control agents to destroy weeds.

▷ **TINGIS CARDUI** is a British lace bug that feeds on thistles. The body may look pale grey, due to a covering of light, powdery wax.

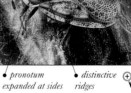

pronotum expanded at sides

distinctive ridges

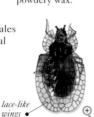

lace-like wings

◁ **DEREPHYSIA FOLIACEA**, with its typical lacy wings, favours ivy as a host. It was found in North America for the first time in 1987.

Length 2–5mm (¹⁄₁₆–³⁄₁₆in)	Feeding habits

| Order HEMIPTERA | Family APHROPHORIDAE | No. of species 850 |

SPITTLE BUGS

Similar to frog-hoppers (below), these bugs are also strong jumpers. Many species are pale to dark brown, with mottling and bands of colour.
• LIFE-CYCLE Eggs are laid in slits in plant tissue. True to their name, the pale, soft nymphs live under a frothy, protective covering that they make by mixing a special fluid with air and expelling it from their anus.
• OCCURRENCE Worldwide. On low-growing vegetation, shrubs, and trees.

eyes wider than they are long •

• bands of colour

• head as wide as pronotum

• hind tibiae have one or two strong spines and a circle of smaller ones

APHROPHORA ALNI is a common species throughout Europe. It is found on various herbaceous plants.

BALSANA SUBFASCIATA comes from Brazil and has characteristic bands of colour on its wings.

| Length 0.6–1.2cm (¼–½in) | Feeding habits |

| Order HEMIPTERA | Family CERCOPIDAE | No. of species 2,400 |

FROG-HOPPERS

As their common name suggests, these squat, round-eyed bugs are highly effective jumpers. Many species have drab colouring but some are brightly coloured red and black or yellow and black. Frog-hoppers look similar to spittle bugs (above) and can also be confused with leaf-hoppers (see p.96).
• LIFE-CYCLE Little is known about the life-cycle of many species, although some lay their eggs in or on soil.
• OCCURRENCE Worldwide, especially in warmer regions. On a wide variety of shrubs, trees, and herbaceous plants.

• round eyes

pronotum wider than head •

tapering • abdomen

△ **GYNOPYGOPLAX THEORA** comes from the Philippines and Korea. It is a strong jumper, and its brightly coloured wings help to warn off predators.

◁ **CERCOPIS VULNERATA** nymphs live under the ground, surrounded by a protective frothy mass that they produce. They feed on plant roots.

• squat body

distinctive black-and-red warning coloration •

| Length 0.5–2cm (³⁄₁₆–¾in) | Feeding habits |

Order HEMIPTERA	Family CICADELLIDAE	No. of species 16,000

LEAF-HOPPERS

Leaf-hoppers are generally slender, with broad or triangular heads. Many species are brown or green, although some can be brightly striped or spotted. These bugs jump very well and are characterized by distinctive hind tibiae, which have an angular cross-section and one or more rows of small spines running along their length.
• **LIFE-CYCLE** Leaf-hoppers communicate with mates by making sounds with special abdominal organs. The sounds travel through foliage by making the leaves vibrate. Rows or clusters of eggs are laid under the epidermis of host plants.
• **OCCURRENCE** Worldwide. Almost anywhere with vegetation.
• **REMARK** Leaf-hoppers are pests of vital crops such as rice and maize.

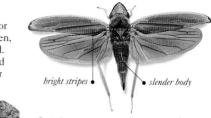

bright stripes • • slender body

△ *GRAPHOCEPHALA COCCINEA* is found on blackberry and ornamental plants in North America and on rhododendron species in Europe.

LEDRA AURITA is a large, flat-bodied leaf-hopper. It has mottled colouring and blends in well against lichen-covered bark.

• *small spines along length of angular tibiae*

Length 0.3–2cm (⅛–¾in), mostly under 1.5cm (⅝in)	Feeding habits 🌿

Order HEMIPTERA	Family CICADIDAE	No. of species 2,500

CICADAS

These bugs have dark brown or green camouflage colouring and a distinctive blunt-headed, tapered shape.
• **LIFE-CYCLE** The males' famous, loud mating songs are produced by a pair of abdominal organs. Females use their ovipositors to cut slits in trees and shrubs, into which they lay their eggs. Hatchlings drop to the ground and burrow into the soil, later emerging and crawling up tree trunks for their final moult. Nymphs may take many years to become fully grown.
• **OCCURRENCE** Worldwide, mainly in warmer regions. On shrubs and trees.

smaller hindwings •

shiny, membranous wings •

△ *ANGAMIANA AETHEREA* comes from India. Like all cicadas, it produces a number of songs, which it uses for courtship and to signal aggression.

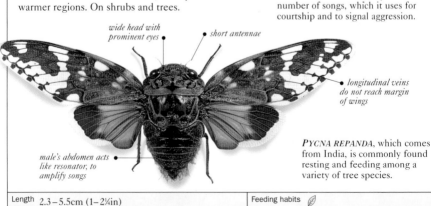

wide head with prominent eyes • • *short antennae*

• *longitudinal veins do not reach margin of wings*

male's abdomen acts like resonator, to amplify songs •

PYCNA REPANDA, which comes from India, is commonly found resting and feeding among a variety of tree species.

Length 2.3–5.5cm (1–2¼in)	Feeding habits 🌿

Order HEMIPTERA	Family DERBIDAE	No. of species 800

DERBIDS

Many derbids are brightly coloured yellow, brown, and pale brown. They have thin legs, a small head, large eyes, and typically long, narrow wings.
• LIFE-CYCLE Little is known, but some species may lay eggs in wood, plant debris, or bark crevices.
• OCCURRENCE Worldwide, especially in tropical and subtropical regions. On trees, flowering plants, and woody fungi on rotting wood.

long wings are held, roof-like, over body at rest

small head

DERBE LONGITUDINALIS has the fragile, moth-like appearance typical of derbids. It is found in Bolivia and Ecuador.

Length 0.7−1.2cm (⁹⁄₃₂−¹⁄₂in)	Feeding habits

Order HEMIPTERA	Family FULGORIDAE	No. of species 800

FULGORID BUGS

The most distinctive feature of many of these bugs is the long head, which may be very strangely shaped. At rest, fulgorids blend into their surroundings; if disturbed, the eye-spots on the hindwings can be flashed to deter predators.
• LIFE-CYCLE Eggs are laid on host plants, surrounded by a protective secretion.
• OCCURRENCE Tropical and subtropical regions. In vegetated habitats.

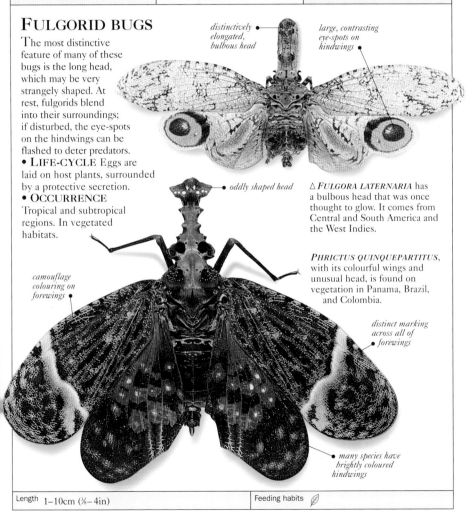

distinctively elongated, bulbous head

large, contrasting eye-spots on hindwings

oddly shaped head

△ **FULGORA LATERNARIA** has a bulbous head that was once thought to glow. It comes from Central and South America and the West Indies.

PHRICTUS QUINQUEPARTITUS, with its colourful wings and unusual head, is found on vegetation in Panama, Brazil, and Colombia.

camouflage colouring on forewings

distinct marking across all of forewings

many species have brightly coloured hindwings

Length 1−10cm (³⁄₈−4in)	Feeding habits

Order HEMIPTERA	Family MEMBRACIDAE	No. of species 2,500

TREEHOPPERS

Named after the fact that they live in trees, and also known as thorn bugs, these insects are mostly green, brown, or black. Some, however, are brightly coloured. They are distinguished from other bugs by the shape of their pronotum. This varies from a thorn or spine, which makes them difficult for a predator to eat, to a large and complex structure that may act as an effective disguise. Nymphs do not have an enlarged pronotum, but may have dorsal spines or lateral expansions.

• LIFE-CYCLE Treehopper eggs are deposited inside plant tissue, and the young go through five nymphal stages before reaching adulthood. Treehoppers feed in groups and suck plant sap. They are often attended by ants who "milk" the nymphs for their carbohydrate-rich excrement (honeydew). In return for the food, the ants guard the treehopper colony.

• OCCURRENCE Worldwide, mainly in warmer areas. On trees in a variety of habitats.

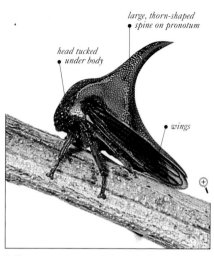

large, thorn-shaped spine on pronotum

head tucked under body

wings

△ *UMBONIA* SPECIES are found in South America, parts of North America, and Southeast Asia. The shape of the pronotum varies but is often spine-like and very sharp – sharp enough, for example, to penetrate shoes and puncture skin.

central ridge of large, spine-like pronotum

lateral spine

▷ *HEMIKYPTHA MARGINATA* is a Brazilian native. Viewed from above (as shown here), only the large, thorn-like pronotum is clearly visible, which deters birds from eating it.

protruding hindlegs

▽ *ANTIANTHE EXPANSA*, from Guatemala, has an effective camouflage device. As seen here, rows of bugs sit feeding head to tail, so that the spines point the same way and make the bugs look like part of a plant.

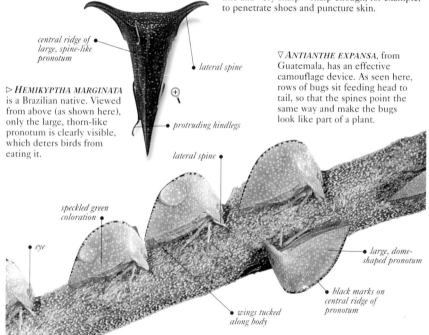

lateral spine

speckled green coloration

eye

large, dome-shaped pronotum

black marks on central ridge of pronotum

wings tucked along body

Length 0.5–1.5cm (³⁄₁₆–⁵⁄₈in)	Feeding habits

| Order HEMIPTERA | Family ALEYRODIDAE | No. of species 1,200 |

WHITEFLIES

These insects resemble tiny moths. The head of these bugs carries a pair of seven-segmented antennae, and the conspicuous wings are either white or mottled, with a distinctive dusting of white, powdery wax over the surface.
• **LIFE-CYCLE** The females lay their eggs on tiny stalks on the undersides of leaves. When the nymphs first hatch out, they move around. However, they lose their legs at the first moult, and after this they become sedentary sap-suckers.
• **OCCURRENCE** Worldwide, especially in warmer regions. On a range of host plants.
• **REMARK** Many whiteflies are serious pests. Well-known examples include *Trialeurodes vaporariorum* (see picture) and *Bemisia tabaci*, which is a widespread pest of cotton and other important crops.

adult, with pale yellow body • *underside of leaf* • *white, powdery wax covers wings* • *nymph* •

TRIALEURODES VAPORARIORUM, the Greenhouse Whitefly, is a widespread pest of cucumbers and tomatoes grown under glass. It may also attack field crops in warm conditions.

| Length 1–3mm (¹⁄₃₂–¹⁄₈in) | Feeding habits |

| Order HEMIPTERA | Family APHIDIDAE | No. of species 2,250 |

COMMON APHIDS

These aphids are small, soft-bodied, and mostly green, pink, black, or brown. The abdomen usually carries a pair of short tubes, called cornicles, from which a substance is secreted to deter predators.
• **LIFE-CYCLE** Females produce large colonies by parthenogenesis (asexual reproduction in which eggs develop without fertilization) and usually give birth to nymphs. Winged adults migrate to a host plant where sap-feeding and parthenogenetic reproduction continues. Later, more winged aphids fly back to the original host plant, where the males and females mate and lay eggs once more.
• **OCCURRENCE** Worldwide. Wherever host plants are found.
• **REMARK** With their huge reproductive potential, aphids are the most destructive of all plant-eating insects. Virtually all crop species are affected by their feeding and by the viral diseases that they transmit.

antennae, with four to six segments •

wingless adult female •

young being born •

eye •

cornicles •

MACROSIPHUM ALBIFRONS, the American Lupin Aphid, is now also found throughout many parts of Europe. This species is a pest and carries diseases such as yellow mosaic virus.

| Length 1–8mm (¹⁄₃₂–⁵⁄₁₆in), most 5mm (³⁄₁₆in) | Feeding habits |

Order HEMIPTERA	Superfamily COCCOIDEA	No. of species 7,000

SCALE INSECTS

Members of this superfamily vary widely in coloration. The sedentary, wingless females are flat and elongate or oval. Their bodies may be covered with waxy secretions that form a soft or scale-like covering. The males, which are uncommon, look very different and may be winged or wingless.

• **LIFE-CYCLE** Nymphs and adult females are sap-suckers. Reproduction can be asexual or sexual, and the reproductive potential of many species is immense. Females lay eggs, commonly on host plants, or give birth to nymphs. Newly hatched nymphs have legs and disperse. Later stages may lose their legs and become sedentary, like the females.

• **OCCURRENCE** Worldwide, especially in subtropical and tropical regions. On host plants in a wide variety of habitats.

• **REMARK** Many species are significant pests of crops such as citrus trees and coffee.

dark, scale-like body

insects clustered around leaf veins

△ *SAISSETIA NIGRA* is a widespread pest of cassava plants and trees and also attacks nutmeg, cinnamon, and teak.

"ground pearls"

△ *MARGARODES* **SPECIES** produce cyst-like nymphs called "ground pearls", which live underground and attack the roots of grasses.

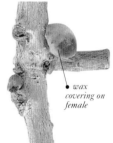

wax covering on female

◁ *PLANOCOCCUS CITRI*, the common Citrus Mealy Bug, also attacks other crops, including coffee, guava, and soya beans.

• mealy bugs cluster under a tree bud

white, powdery covering

△ *CEROCOCCUS QUERCUS* often causes great damage to ornamental trees in North America.

DACTYLOPIUS TOMENTOSUS has been used to control cactus weed in parts of South Africa.

Length 0.1–3cm (¹⁄₃₂–1¼in), most under 1cm (⅜in)	Feeding habits

Order HEMIPTERA	Family PSYLLIDAE	No. of species 1,500

JUMPING PLANT LICE

These variously coloured bugs look like small leaf-hoppers (see p.96), but with longer antennae. The two pairs of oval wings are held, roof-like, over the body. The head has a short, three-segmented beak.

• **LIFE-CYCLE** Females lay stalked eggs on or in plants, and some may cause gall formation or rolling of leaves. The flat nymphs develop wing pads as they get older.

• **OCCURRENCE** Worldwide. On any suitable host plant, in a wide variety of habitats.

• **REMARK** Some species are significant plant pests.

wings held over body when at rest

small black spots on forewings

CACOPSYLLA PYRICOLA, the Pear Psyllid, is a significant pest of pear trees throughout the northern hemisphere.

Length 1.5–5mm (¹⁄₁₆–³⁄₁₆in)	Feeding habits

THRIPS

C OMMONLY KNOWN as thrips, the order Thysanoptera contains 8 families and 5,000 species. They are small, slender insects usually with two pairs of narrow, hair-fringed wings. The head bears short antennae, conspicuous compound eyes, and distinctive sucking mouthparts, which include a pair of mandibles in which one is small and the other is needle-like. There is a sticky, inflatable structure between the tarsal claws that aids grip on smooth surfaces.

Thrips are closely related to bugs but are unusual in that they undergo neither complete nor incomplete metamorphosis. As in members of the order Hymenoptera (see pp.178–206), fertilized eggs produce females and unfertilized eggs produce males. There are also one or more pupa-like stages after the two nymphal stages. The females of some species have a saw-like ovipositor and lay their eggs inside plant tissue, while others lack an ovipositor and lay their eggs in cracks and crevices or on the surface of host plants.

Thrips may be herbivorous or predacious. A few species show simple forms of social behaviour, and in some there are soldiers who defend their colony. Many thrips are plant pests – especially of cereal crops.

Order THYSANOPTERA	Family AEOLOTHRIPIDAE	No. of species 250

BANDED THRIPS

Also called predacious thrips, because of the feeding habits of some species, these thrips are yellow-brown or dark. The body is round in cross-section, and the wings, which lie parallel to each other when folded, often have cross bands or stripes.

• LIFE-CYCLE Although most banded thrips feed on other small insects, including other thrips, some species feed on pollen grains. The females of most species lay eggs inside host plants with a saw-like ovipositor. When the yellow or orange nymphs are fully grown, they form a silken cocoon underground.

• OCCURRENCE Worldwide, especially in temperate regions. On host plants, and especially on flowers, in various habitats.

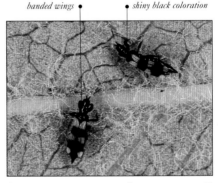

banded wings •   • *shiny black coloration*

AEOLOTHRIPS SPECIES are often recognized by their banded wings. Most of the species in this genus are found only in the northern hemisphere.

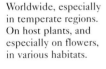

fringes of hairs on wings •

• *distinctive blunt head*

• *mottled abdomen*

AEDOTHRIPS TENUICORNIS is common throughout Europe, especially among the yellow flowers of legumes and composites.

Length 1–2mm (¹⁄₃₂–¹⁄₁₆in) (body length)	Feeding habits

Order THYSANOPTERA	Family PHLAEOTHRIPIDAE	No. of species 2,700

TUBE-TAILED THRIPS

Members of this family have larger, stouter bodies than most thrips and the abdomen has a tubular, pointed end. Most species are dark but often have light or mottled wings. When present and folded, the wings overlap each other.
• LIFE-CYCLE Eggs are laid in cracks and crevices. Most species eat fungi, while some feed inside galls, on plants, or in decaying wood. A few eat mites and small insects. The nymphs are usually red or yellowish, feed alongside the adults in groups, and communicate by sounds.
• OCCURRENCE Worldwide, mainly in tropical and subtropical areas. On herbaceous plants, trees, and shrubs; in soil and leaf-litter.
• REMARK Some species are crop pests.

last abdominal segment cylindrical and tapered • *developing wings* •

PHLAEOTHRIPIDAE SPECIES feed on plants. This specimen is seen resting on a bromeliad plant in South America. It will soon moult to the adult stage and develop full-sized wings.

Length 0.1–1.2cm (¹/₃₂–¹/₂in), most under 5mm (³/₁₆in)	Feeding habits ✱ ⬭ 🍄

Order THYSANOPTERA	Family THRIPIDAE	No. of species 1,750

COMMON THRIPS

The coloration of these flat thrips varies from pale yellow to brown or black. The hair-fringed wings are very narrow, pointed at the ends, and sometimes banded. Females have a saw-like ovipositor that bends downwards.
• LIFE-CYCLE Reproduction can occur asexually, and the females use their saw-like ovipositors to insert their eggs inside plants or flowers. The adults and nymphs of most species suck plant juices, although some species eat fungi or even suck the juices of other insects. When fully grown, the nymphs enter a pupa-like stage either on the plant or in the soil.
• OCCURRENCE Worldwide. On the leaves and flowers of a huge range of host plants.
• REMARK Many common thrips are serious pests of a wide range of crops, including tobacco, cotton, and beans.

eyes with large facets • *flat, dark body* • *dark hairs on body* •

THRIPS FUSCIPENNIS is a dark-coloured species with distinctive hairs on its body. Widespread across the northern hemisphere, these thrips are found inside a wide range of plant species.

third segment of eight-segmented antennae is pale •

THRIPS SIMPLEX, the Gladiolus Thrips, originally came from South Africa. It has now spread much further afield and is found wherever gladioli flowers are grown.

• *shiny, dark body*

• *clear, hair-fringed wings*

• *pale tarsi*

Length 0.7–2mm (¹/₃₂–¹/₁₆in)	Feeding habits ✱ ⬭ 🍄

ALDERFLIES AND DOBSONFLIES

T̲HE ORDER MEGALOPTERA is relatively small. It is divided into 2 families and 300 species. They are the most primitive insects that develop by complete metamorphosis. There are two distinct families: the alderflies (Sialidae) and the dobsonflies (Corydalidae).

Both families have soft bodies and are drably coloured with two pairs of large wings of almost equal size. When the wings are folded, they are held, roof-like, over the body. Weak fliers, species in this order never move far from water.

Adults may have large mandibles, but they do not feed. Their larvae, which are aquatic and have abdominal gills, are predacious and eat anything they can kill. After up to 11 larval stages, pupation occurs inside a chamber made by the larva in sand, soil, or moss.

Order MEGALOPTERA	Family CORYDALIDAE	No. of species 200

DOBSONFLIES

The wings of these usually large, soft-bodied insects are clear or have grey or brown areas. Unlike alderflies, they have three ocelli. The males may have huge mandibles, used for male-to-male combat or for holding the female.
• **LIFE-CYCLE** Eggs are laid near water. The bottom-dwelling larvae have eight pairs of simple filaments with basal gill tufts and may take years to mature.
• **OCCURRENCE** Worldwide, especially in temperate regions. In running water.

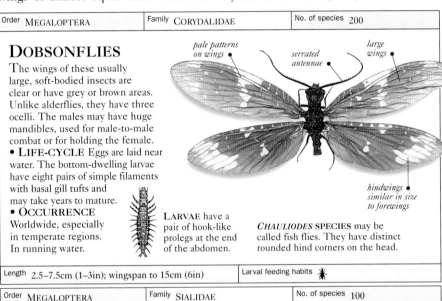

pale patterns on wings •
serrated antennae •
large wings •
hindwings • similar in size to forewings

LARVAE have a pair of hook-like prolegs at the end of the abdomen.

CHAULIODES SPECIES may be called fish flies. They have distinct rounded hind corners on the head.

Length 2.5–7.5cm (1–3in); wingspan to 15cm (6in)	Larval feeding habits

Order MEGALOPTERA	Family SIALIDAE	No. of species 100

ALDERFLIES

These dark, smoky-winged species are much smaller than dobsonflies (see above) and have no ocelli. Adults are often found resting on waterside vegetation.
• **LIFE-CYCLE** Females lay large egg masses near water, into which the hatched larvae drop. These aquatic larvae take one year to mature. They have seven pairs of feathery, lateral gills and a single tail filament at the end of the abdomen.
• **OCCURRENCE** Worldwide, especially in temperate regions. In ponds, canals, streams, and slow-moving water.

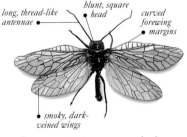

long, thread-like antennae •
blunt, square head •
curved forewing margins •
smoky, dark-veined wings •

LARVAE have well-developed legs and are good crawlers.

SIALIS SPECIES are common in the northern hemisphere. Fully grown larvae crawl out of the water and pupate in damp soil above the water margin.

Length 1–2cm (³⁄₈–³⁄₄in), most under 1.5cm (⁵⁄₈in)	Larval feeding habits ✳

SNAKEFLIES

R APHIDIIDAE and Inocellidae are the two families that make up the order Raphidioptera, which contains 150 species in total. Both families have similar features. All snakeflies have two pairs of wings and a slightly flattened head with forward-pointing mouthparts that are used for chewing. The pronotum is typically elongate. Species belonging to the Inocellidae, the smaller of the two families, are distinguishable from the Raphidiidae in that they do not have ocelli and their antennae are long. Snakeflies are closely related to alderflies (see p.103), but their larvae are terrestrial and they do not have gills.

Snakeflies live in woodlands where there is a plentiful supply of vegetation. Both the adults and their larvae are predacious but they also scavenge for a significant amount of their food.

During mating, the male is positioned underneath the female. Several hundred eggs may be laid, in groups of up to 100, either in tree bark or in rotten wood. Metamorphosis is complete.

The name snakefly refers to the snake-like way in which the adults catch hold of their prey. They do this by raising up their head, at the end of its elongate prothorax, and moving it forward to seize the food.

Order RAPHIDIOPTERA	Family RAPHIDIIDAE	No. of species 85

SNAKEFLIES

These dark-coloured insects have a distinctive neck consisting of an elongated pronotum on which the head can be raised. The head is typically broad across the eyes and tapering towards the rear. Female snakeflies are slightly larger than the males.
• **LIFE-CYCLE** Females use a long, slender ovipositor to lay eggs in slits in bark. The elongate larvae can be found under loose bark, in decaying tree stumps, and among leaf-litter. Like their parents, the larvae feed mostly on beetle larvae and other soft-bodied insects.
• **OCCURRENCE** Primarily in the northern hemisphere. In woods among vegetation.

LARVAE are flat, with a square head, small eyes, short, curved jaws, and strong legs.

slightly flat head •   *characteristic neck with elongated pronotum* •   *long ovipositor* •

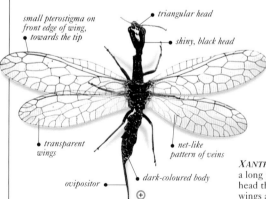

small pterostigma on front edge of wing, • *towards the tip*

• *triangular head*

• *shiny, black head*

• *transparent wings*

• *net-like pattern of veins*

ovipositor •

• *dark-coloured body*

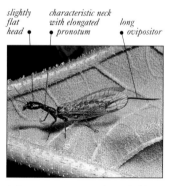

△ *AGULLA* SPECIES are found in North America and Canada, from the Rocky Mountains west through the Great Basin to the Pacific coast.

XANTHOSTIGMA XANTHOSTIGMA has a long pronotum and a distinct, broad head that tapers to the rear. The four wings are of almost equal size.

Wingspan 0.6–3cm (¼–1¼in)	Larval feeding habits

ANTLIONS, LACEWINGS, AND THEIR RELATIVES

T HE ORDER NEUROPTERA includes 17 families and 4,000 species. Its members generally have large compound eyes, chewing mouthparts, and antennae that are usually longer than the head and thorax combined. They also have two pairs of equally sized wings, which are held roof-like over the body when not in use. The major wing veins are forked or twigged near the margins.

The adults in this order are mostly predatory, but a few feed on pollen and nectar. Most species hunt either in the evening or after dark.

Metamorphosis is complete. The larvae have curved mouthparts that form a hollow tube, through which the juices of prey are sucked up. After three nymphal stages, larvae pupate inside a fragile silk cocoon.

Order NEUROPTERA	Family ASCALAPHIDAE	No. of species 450

OWLFLIES

Large, conspicuous, and sometimes with highly patterned wings, these species are active mostly after dark and are often attracted to lights. The grey, black, or red-brown body is elongate, and the antennae have clubbed ends. The wings are pale to smoky, with yellow or darker markings. Adults are agile hunters, actively chasing and seizing prey in the air.
• **LIFE-CYCLE** Females lay rows or spirals of up to 50 eggs on twigs or grass stems. The larvae wait on the ground, in leaf-litter, or on tree trunks for suitably sized prey to come along.
• **OCCURRENCE** Worldwide, especially in warm regions. In grassland or warm, dry woodland.
• **REMARK** Some owlflies look very similar to dragonflies (see pp.53–55) although the resemblance is superficial.

LARVAE are oval and flat, with expansions at the sides of the abdomen. Some have jaws that open very wide.

LIBELLOIDES COCCAJUS is a European species with a distinctive wing shape and a large area of black at the bases of the hindwings.

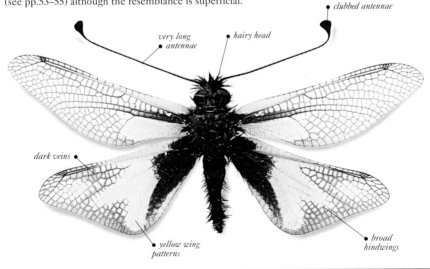

• *clubbed antennae*

very long antennae

• *hairy head*

dark veins •

• *yellow wing patterns*

• *broad hindwings*

Wingspan 3–12cm (1¼–4¾in)	Larval feeding habits

Order NEUROPTERA	Family CHRYSOPIDAE	No. of species 1,600

COMMON LACEWINGS

Although some species are brown, these insects are generally green. The wings are iridescent, with veins that form complex patterns and fork at the wing margins. The eyes have a bright golden or reddish shine. Adults are nocturnal and are attracted to lights, often entering houses to hibernate. Many have special bat-detecting sensors in their wings.

• **LIFE-CYCLE** Females lay stalked eggs on vegetation, and the pale larvae pupate in round silk cocoons stuck to leaves. Many larvae cover themselves with the bodies of prey as a disguise. Adults and larvae are predators of aphids, thrips, scale insects, and mites.

• **OCCURRENCE** Worldwide. On vegetation in varied habitats, including arid areas, and in ants' nests.

veins forked at hind margin

long, thread-like antennae

CHRYSOPA SPECIES are delicate and large-winged insects. When at rest, they hold their wings, roof-like, over their body.

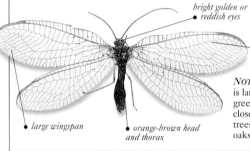

bright golden or reddish eyes

two zig-zag veins in outer half of wings

large wingspan

orange-brown head and thorax

NOTHOCHRYSA CAPITATA is larger and duller than the green lacewings. It is found close to the trunks of various trees and in the crowns of oaks and pines.

LARVAE have curved jaws, hairy warts, and well-developed legs.

Wingspan 1–5cm (⅜–2in)	Larval feeding habits 🐜

Order NEUROPTERA	Family MANTISPIDAE	No. of species 300

MANTISPIDS

Also called mantidflies, mantispids have front legs exactly like those of praying mantids (see p.73), which they use to seize prey. The first segment of the thorax is elongate, and the two pairs of narrow wings are of roughly equal size.

• **LIFE-CYCLE** Small, white, short-stalked eggs are laid in groups of several hundred on tree bark. The young nymphs are mobile and hunt for spiders' egg sacs, inside which they feed. Some species parasitize bees.

• **OCCURRENCE** Worldwide, mainly in warm, temperate and tropical regions. In well-vegetated areas.

wings held in wasp-like position

wasp-like "waist"

black-and-yellow markings

LARVAE have three pairs of thoracic legs. They become maggot-like as they mature.

CLIMACIELLA SPECIES have a distinctive body shape. The one shown here is probably protected from attack by its bright, wasp-like coloration.

Wingspan 1–5.5cm (⅜–2¼in)	Larval feeding habits 🐜 🕷

Order NEUROPTERA	Family MYRMELEONTIDAE	No. of species 1,000

ANTLIONS

These large, soft, slender insects resemble damselflies (see pp.51–53). The head is broader than the pronotum, with large, conspicuous eyes and club-ended antennae that are about as long as the head and thorax together. The long, narrow wings may have brown or black patterns.

• **LIFE-CYCLE** Eggs are laid in soil or sand, singly or in small groups. The larvae eat insects and spiders. Some antlions construct conical pits to trap prey. The larvae live in these pits, with only their sharp, spiny mandibles showing, and flick sand grains at prey to knock them into the lair. Other larvae live on tree trunks, in soil and debris, or under stones.

• **OCCURRENCE** Worldwide, especially in semi-arid areas in subtropical and tropical regions. In open woodland, scrub grassland, and dry, sandy areas.

• **REMARK** Antlion larvae are sometimes called doodlebugs.

LARVAE have large, curved, toothed jaws and long legs. The large abdomen tapers towards the rear.

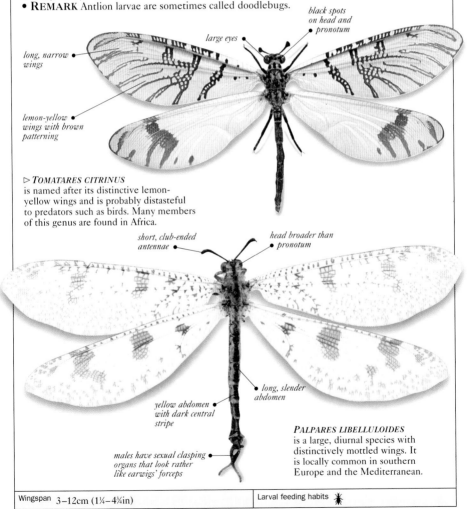

large eyes

black spots on head and pronotum

long, narrow wings

lemon-yellow wings with brown patterning

▷ *TOMATARES CITRINUS*
is named after its distinctive lemon-yellow wings and is probably distasteful to predators such as birds. Many members of this genus are found in Africa.

short, club-ended antennae

head broader than pronotum

long, slender abdomen

yellow abdomen with dark central stripe

males have sexual clasping organs that look rather like earwigs' forceps

PALPARES LIBELLULOIDES
is a large, diurnal species with distinctively mottled wings. It is locally common in southern Europe and the Mediterranean.

Wingspan 3–12cm (1¼–4¾in)	Larval feeding habits 🐜

Order NEUROPTERA	Family NEMOPTERIDAE	No. of species 150

THREAD-WINGED LACEWINGS

The forewings of these species are usually large and patterned, whereas the hindwings are either long and slender, with rounded ends, or long and thread-like.
• LIFE-CYCLE Eggs are laid on the ground, in dry soil or sand under rock outcrops and cave entrances. The larvae develop *in situ*, and pupation occurs in a silk cocoon.
• OCCURRENCE In subtropical and tropical areas worldwide except North America. On soil or sandy ground.

LARVAE of some species have very elongated necks.

large forewings with zig-zag markings

• very long, thread-like hindwings

NEMOPTERA SINUATA is native to southern Europe. It has brown forewing markings and brown bands at the end of its hindwings.

Wingspan 2–8cm (¾–3¼in)	Larval feeding habits 🦎

Order NEUROPTERA	Family OSMYLIDAE	No. of species 150

OSMYLIDS

These slender lacewings have broad, patterned wings and thread-like antennae.
• LIFE-CYCLE Rows of eggs are laid on tree trunks or foliage near water. Some larvae are semi-aquatic (living in wet moss, for example) and eat insect larvae; others live under tree bark.
• OCCURRENCE Worldwide, except North America. In wooded areas near to a steady flow of water, such as a stream.

LARVAE have straight, needle-like mouthparts.

OSMYLUS FULVICEPHALUS is widespread in Europe. Like all osmylids, it is a weak flier, despite its large wings.

thread-like antennae

broad, spotted wings

Wingspan 1.4–3cm (⅝–1¼in)	Larval feeding habits 🦎

Order NEUROPTERA	Family SISYRIDAE	No. of species 50

SPONGEFLIES

Also known as spongillaflies, these small insects have slender antennae and lacy wings. Adults do not fly very far from water and are active at twilight and after dark.
• LIFE-CYCLE Eggs are laid singly or in small groups on overhangs, and hatched larvae drop into the water. The larvae have very elongate, slender jaws for sucking out the body contents of sponges.
• OCCURRENCE Worldwide. Near a source of fresh water, wherever sponges occur.

large, lacy wings

long, dark antennae

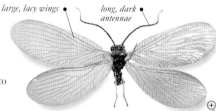

LARVAE have three pairs of legs and abdominal gills.

SISYRA FUSCATA is widespread across Europe. The adults catch and feed on aphids and are often attracted to lights after dark.

Wingspan 0.6–1.2cm (¼–½in)	Larval feeding habits 🦎 🦎

BEETLES

T HERE ARE 166 families and 370,000 species in the order Coleoptera. About one in three insects in existence today is a beetle, and they have successfully colonized every sort of terrestrial and freshwater habitat. They range from tiny insects less than 1mm (1/32in) in length to tropical giants measuring 18cm (7in).

Although beetles vary enormously in both their shape and their coloration, a major distinguishing feature is their toughened forewings, also known as elytra. These hard forewings protect the larger, membranous hindwings that are folded underneath. The elytra may be short, but in all species they meet down the middle of the body. It is the possession of protective elytra and a compact, strong body that has helped beetles become successful, allowing them to dig or squeeze themselves into all kinds of spaces and survive in a huge variety of habitats. In aquatic species, the space beneath the elytra provides a valuable storage area for air.

Mating usually takes place with the male clinging to the female's back. Most beetle species are herbivorous, but there are many scavengers and predators, as well as a few specialized parasitic species. Metamorphosis is complete.

Order COLEOPTERA	Family ANOBIIDAE	No. of species 1,500

WOODWORM

The larvae of some of these small beetles attack wood – hence the common name. They are pale brown to black and vary from elongate to oval. The head is often hooded by the pronotum, and the short legs fit into grooves on the underside of the body. The last three segments of the antennae are lengthened or expanded.

• LIFE-CYCLE Eggs are laid on a suitable food source, and the larvae bore inside, making circular tunnels. Pupation takes place just below the surface.
• OCCURRENCE Worldwide. In woods, stores, warehouses, and other buildings.
• REMARK This family includes the notorious Furniture Beetle or Common Woodworm (*Anobium punctatum*).

branched antennae in males

short legs

elongate shape

△ *XESTOBIUM RUFOVILLOSUM*, the Deathwatch Beetle, is so-named because its mating calls (head-taps against tunnels in wood) were once commonly heard in quiet rooms during wakes.

△ *PTILINUS PECTINICORNIS* is found all over central Europe. It sometimes becomes a pest by attacking wooden furniture.

antennae have 8–11 segments

PTINOMORPHUS IMPERIALIS is common in central Europe. It may be found on hawthorn flowers in springtime.

distinctive patterns on elytra

LARVAE that eat wood contain symbiotic yeasts that break cellulose down into sugars.

Length 2–9mm (1/16–11/32in), mostly 2–6mm (1/16–1/4in)	Larval feeding habits

Order COLEOPTERA	Family ANTHRIBIDAE	No. of species 3,000

FUNGUS WEEVILS

Most of these weevils are oblong. They are usually dark and densely covered with patterns of white or dark scales or hairs. The antennae usually have 11 segments.
• LIFE-CYCLE Eggs are laid, and larvae develop, inside rotten wood, fungi, or plant tissue. Adults mostly eat fungi.
• OCCURRENCE Worldwide, especially in tropical regions. On twigs, under bark, and in fungi, wood, and seed pods.

LARVAE are pale and curved, often with short, two-segmented legs.

broad nose

MECOCERUS GAZELLA is found in Southeast Asia. It has exceptionally long antennae.

very long antennae, with 11 segments

Length 0.05 – 3.8cm (½–1½in), most under 2cm (¾in)	Larval feeding habits

Order COLEOPTERA	Family BOSTRICHIDAE	No. of species 700

BRANCH-BORING BEETLES

These cylindrical beetles are mostly dark brown or black in colour. The head points downwards.
• LIFE-CYCLE Eggs are laid under the surface of wood. Adults and larvae may tunnel into wood, branches, or twigs. Their guts contain micro-organisms that help them to digest wood.
• OCCURRENCE Worldwide, especially in tropics and subtropics. Inside wood of host plants.

BOSTRICHUS CAPUCINUS is a widespread European species.

parallel-sided, cylindrical body

LARVAE are pale and short-haired.

Length 0.3 – 5cm (⅛–2in), most 0.6 – 2cm (¼–¾in)	Larval feeding habits

Order COLEOPTERA	Family BRENTIDAE	No. of species 2,500

PRIMITIVE WEEVILS

These parallel-sided, elongate beetles are typically black, brown, or yellow. The long, narrow head has a distinctive rostrum, bearing thread- or bead-like antennae with 11 segments.
• LIFE-CYCLE Using her rostrum, the female beetle bores a hole into wood, where she deposits her eggs. The larvae tunnel into the sapwood of dead or decaying trees and may feed on fungi. The adults feed on fungi and sap.
• OCCURRENCE Tropical regions. In forests.

BRENTHUS SPECIES are usually hairless. Like all members of this family, they are closely related to weevils (see p.117).

mandibles

elongate rostrum

eyes

antennae arise part-way down head

elongate body, with parallel sides

LARVAE are elongate and have short, one-segmented legs.

Length 0.3 – 8.6cm (⅛–3¼in)	Larval feeding habits

Order COLEOPTERA	Family BUPRESTIDAE	No. of species 15,000

JEWEL BEETLES

Also called metallic wood-boring beetles, most jewel beetles are a brilliant, metallic green, red, or blue, with stripes, bands, and spots. Typically they are slightly flat, tapering towards the rear, with large eyes and short antennae.
• LIFE-CYCLE Eggs are laid in wood. The larvae of most species chew oval tunnels into dead or dying trees. Adults feed on flowers, nectar, and pollen.
• OCCURRENCE Worldwide, primarily in tropical regions. In woods and forests.
• REMARK Some species have heat sensors at the base of the middle legs that detect freshly burned forest – good mating and egg-laying sites. Many species are timber pests.

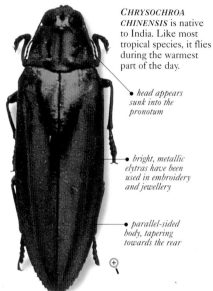

CHRYSOCHROA CHINENSIS is native to India. Like most tropical species, it flies during the warmest part of the day.

• *head appears sunk into the pronotum*

• *bright, metallic elytras have been used in embroidery and jewellery*

• *parallel-sided body, tapering towards the rear*

• *hairy body surface is characteristic of this species*

JULODIS KLUGII, from South Africa, has distinctive hairs all over its body, although the majority of jewel beetles are smooth and shiny, with pits or striations.

LARVAE are pale, with a large, expanded prothorax and a tapering abdomen, giving rise to another common name – flat-head borers.

Length 0.2–6.5cm (¹⁄₁₆–2½in); most under 3cm (1¼in)	Larval feeding habits

Order COLEOPTERA	Family CANTHARIDAE	No. of species 4,500

SOLDIER BEETLES

Most soldier beetles have soft, elongate bodies with parallel sides. The head has distinctive curved jaws and thread-like antennae. Although they are predacious, some species will also feed on pollen and nectar.
• LIFE-CYCLE Eggs are scattered on the ground. Most larvae hunt for prey in soil, decaying timber, and leaf-litter, and under bark. The larvae of a few species eat plant material.
• OCCURRENCE Worldwide. On flowers and other vegetation in hedgerows, meadows, and woodland margins.
• REMARK The common name comes from the red, yellow, and black colouring of many species, reminiscent of old military uniforms.

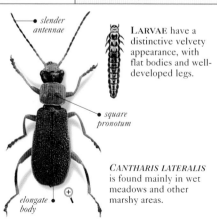

• *slender antennae*

LARVAE have a distinctive velvety appearance, with flat bodies and well-developed legs.

• *square pronotum*

CANTHARIS LATERALIS is found mainly in wet meadows and other marshy areas.

elongate body

Length 0.3–3cm (⅛–1¼in)	Larval feeding habits

Order COLEOPTERA	Family CARABIDAE	No. of species 29,000

GROUND BEETLES

These long, slightly flat beetles may be dull or shiny and are usually brown or black, often with a metallic sheen. The head, thorax, and abdomen tend to be clearly differentiated, and the elytra usually have obvious striations. Most species are nocturnal hunters.

• **LIFE-CYCLE** Eggs are laid on the ground and on vegetation and decaying wood and fungi. Like the adults, larvae are mainly predacious but will eat carrion; a few species are partly herbivorous.

• **OCCURRENCE** Worldwide. On the ground, under stones and logs, and among debris and leaf-litter. Some species live in the foliage of shrubs and trees.

• **REMARK** A few species deter predators with blasts of hot, caustic substances that they expel, with an audible "pop", from the end of the abdomen.

huge, slicing, sickle-shaped mandibles

yellow spots on expansions of the pronotum

△ *ANTHIA THORACICA* is a ground-living predator. Like all members of the genus *Anthia*, it can produce defensive chemicals that it sprays at attackers.

LARVAE are elongate and black or dark brown with well-developed legs and strong jaws.

flat elytra let beetle squeeze under bracket fungi and tree bark

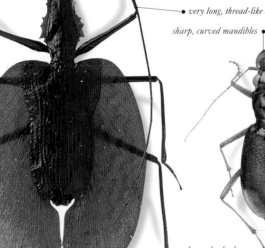

extremely long head

very long, thread-like antennae

sharp, curved mandibles

long, slender legs for fast running

outline of head and body resembles a violin – hence the common name

MORMOLYCE PHYLLODES, or the Violin Beetle, lives in the forests of Southeast Asia and feeds on insect larvae and snails.

MEGACEPHALA AUSTRALIS has bright, metallic coloration. Despite this, it hunts after dark, like all species in the genus *Megacephala*.

Length 0.2–8cm (¹⁄₁₆–3¼in)	Larval feeding habits

Order COLEOPTERA	Family CERAMBYCIDAE	No. of species 30,000

LONGHORN BEETLES

Also known as timber beetles, members of
this family have elongate, parallel-sided bodies
and long antennae – up to four times their
body length. Their colouring is very varied.
• **LIFE-CYCLE** Single eggs are laid on plants
and trees. Most larvae eat dead or decaying
wood, aided by internal micro-organisms that
help them digest cellulose, but some attack
live trees or bore into stems or seeds. Adults
may eat pollen, nectar, leaves, or sap.
• **OCCURRENCE** Worldwide, mainly in
tropical regions. In forests and woodland.
• **REMARK** Many species are serious pests
of timber, fruit, and ornamental trees.

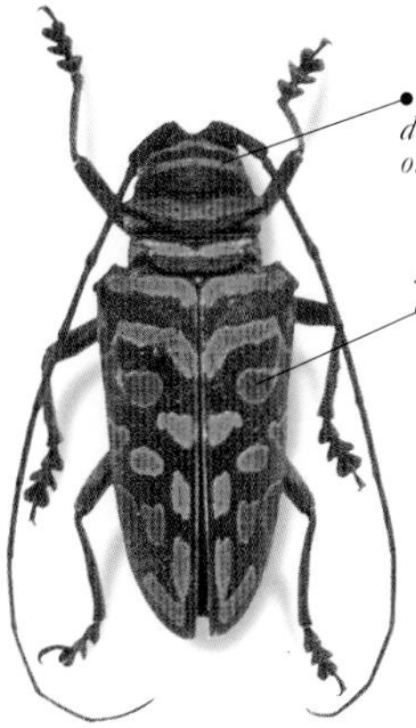

head has
distinctive
orange stripes

surface patterned with
spots and bands

◁ *STERNOTOMIS
BOHEMANNI* has the
typical longhorn
shape. The females
use their mandibles
to cut through bark,
under which they
lay their eggs.

third segment of tarsus
has two lobes, as with
all family members

strong, toothed
mandibles

eyes notched
where antennae
arise

LARVAE are
cylindrical in
shape and may
take months, or
years, to develop.

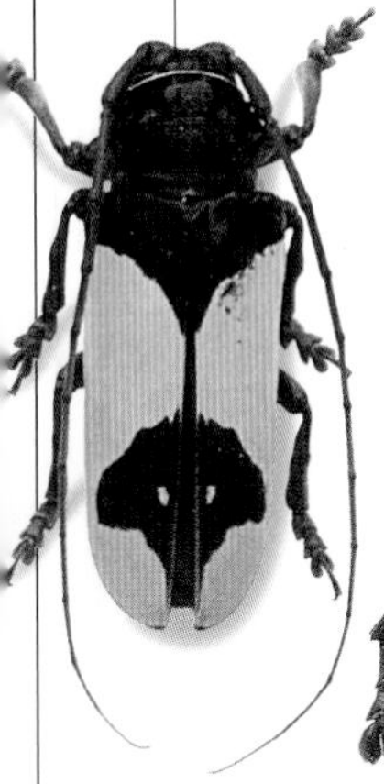

mandibles point
downwards

pronotum has
spiny margin

*PHOSPHORUS
JANSONI* is
found in Africa. It lays eggs in
trees, and its larvae attack some
economically important species
such as the cola tree.

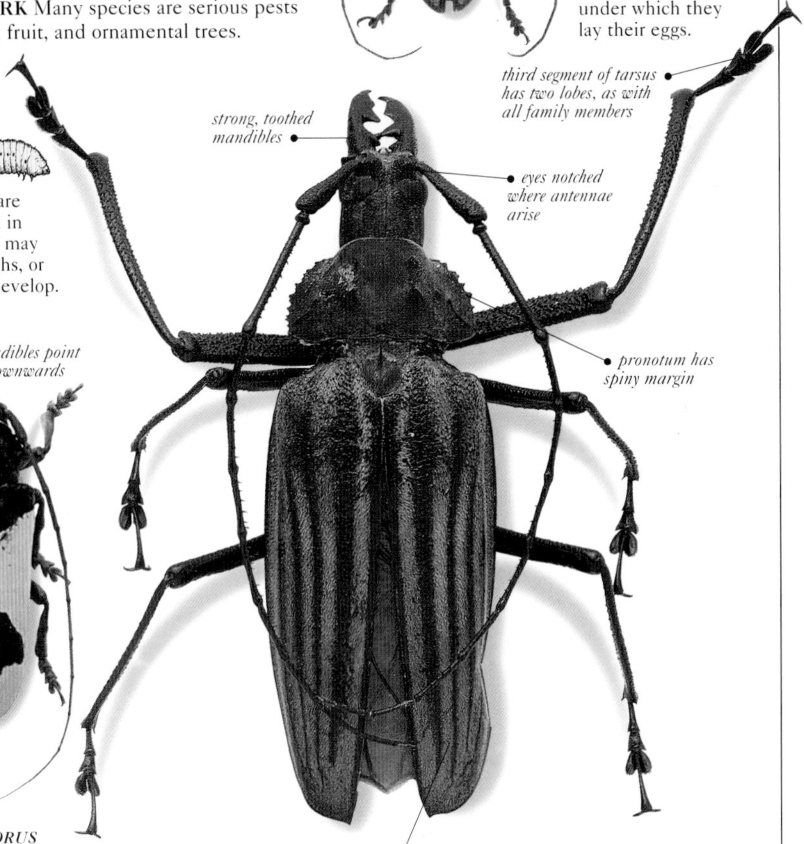

parallel-sided body

XIXUTHRUS HEROS males are
territorial and fight with each
other for control of suitable
egg-laying sites.

Length 0.3–15cm (⅛–6in), most under 4.5cm (1¾in)	Larval feeding habits

Order COLEOPTERA	Family CETONIIDAE	No. of species 3,500

FLOWER CHAFERS

Also called fruit chafers, many of these robust, squarish, slightly flat beetles are brightly coloured and often shiny. The head may have projections of varying lengths, which are usually more developed in males.
• **LIFE-CYCLE** Eggs are laid in decaying plant matter, dry carcasses, and the ground. The larvae eat rotting plant material, dung, and wood; the adults mostly eat plant sap, pollen, and fruit.
• **OCCURRENCE** Worldwide, especially in tropical and subtropical regions. In any well-vegetated habitat.

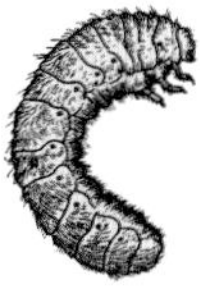

strong claws

head projection

shiny surface

LARVAE are C-shaped, and can wriggle along on their backs when exposed.

▷ *DICRONORHINA DERBYANA* specimens may be less striped than the one shown here, with larger green areas.

large antennal club

distinctive striped elytra

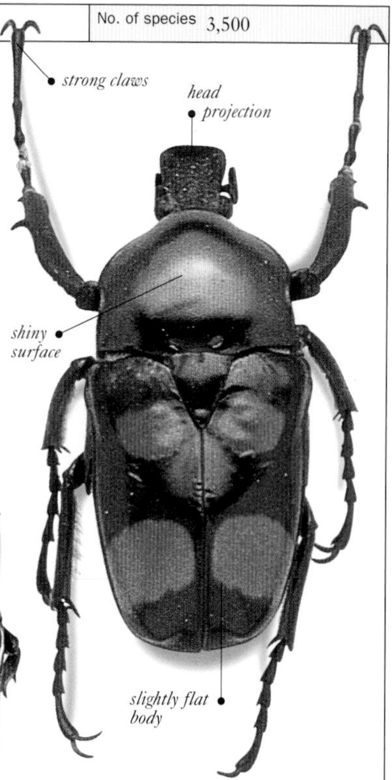

slightly flat body

△ *IUMNOS RUCKERI* comes from northern India and Burma. Its squarish outline, shiny surface, and slightly flat shape are characteristic of flower chafers.

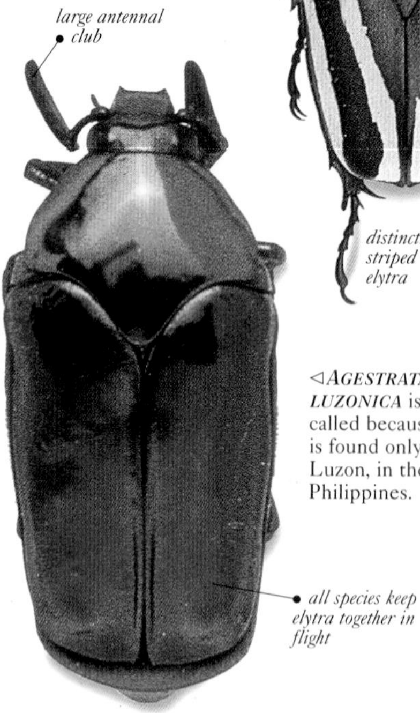

◁ *AGESTRATA LUZONICA* is so-called because it is found only in Luzon, in the Philippines.

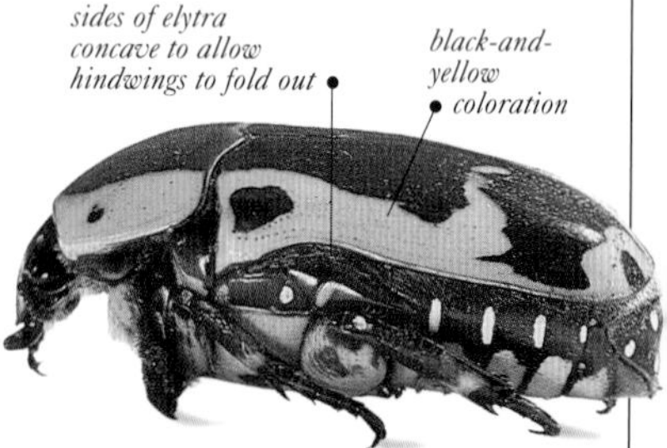

sides of elytra concave to allow hindwings to fold out

black-and-yellow coloration

all species keep elytra together in flight

PACHNODA SINUATA is a stout species from South Africa. It has several forms, in varying patterns of yellow and black. There are more species of this genus in tropical Africa than anywhere else.

Length 1–7cm (⅜–2¾in)	Larval feeding habits

Order COLEOPTERA	Family CHRYSOMELIDAE	No. of species 35,000

LEAF BEETLES

These beetles range from cylindrical and elongate to round-backed. Many species are brightly coloured or metallic. The antennae are less than half the length of the body, which is typically smooth and hairless.
• **LIFE-CYCLE** Groups of eggs are laid on host plants or in soil. The larvae usually feed externally on stems, foliage, and roots. Pupation normally occurs in the soil.
• **OCCURRENCE** Worldwide. Widespread on plants in all terrestrial habitats.
• **REMARK** Many species are pests: the brightly striped Colorado Beetle (*Leptinotarsa decemlineata*, see below) eats potatoes, tomatoes, and aubergines.

LARVAE are long and slightly curved, with well-developed legs.

antennae less than half length of body

smooth, hairless body

▷ *EUMOLPUS* **SPECIES** are found in tropical parts of South America and are typical of this family in appearance.

robust antennae

squarish pronotum characteristic of Sagra species

stout femora

curved tibiae

△ *SAGRA* **SPECIES** may be called jewelled frog beetles due to their strong hindlegs, used in male-to-male combat and for defence. The larvae live inside plant stems.

head hidden by pronotum

broad tarsal segments for good grip on leaf surface

△ *CALISPIDEA REGALIS* is native to Brazil. This very round species has red, net-like patterning and flat expansions of the elytra at the sides.

convex oval body

black and yellow to orange-red longitudinal stripes

▷ *LEPTINOTARSA DECEMLINEATA*, the Colorado Beetle, is found in Asia, North America, and Europe. It is a pest of various food crops, including potatoes.

Length 0.1–3cm (½–1¼in)	Larval feeding habits

Order COLEOPTERA	Family CLERIDAE	No. of species 3,500

CHEQUERED BEETLES

Most species in this family have soft, elongate, slightly
flat bodies. They are often covered with long hairs,
and many have distinctive clubbed antennae. Many
species are brightly coloured red, green, blue, or pink,
and a few have brown or black camouflage colouring.
• LIFE-CYCLE The larvae of these beetles typically
prey on the larvae of wood-boring beetles, and so eggs
are frequently laid in dead wood where the hosts are
burrowing. The larvae of some species in this family
eat bee and wasp larvae and grasshopper eggs.
• OCCURRENCE Worldwide, mainly
in subtropical and tropical regions. On tree
trunks, flowers, and foliage; in fungi and leaf-litter;
and in the tunnels of wood-boring beetles and the
nests of social insects such as bees and termites.

clubbed antennae

long hairs

purple metallic sheen

elongate body with parallel sides

heart-shaped second, third, and fourth tarsal segments

bright, chequered, wasp-like colouring

LARVAE often have
projections at the
end of the abdomen.

TRICHODES CRABRONIFORMIS
comes from central Europe. Its
larvae develop in the nests of
certain bee species.

Length 0.2–5cm (¹⁄₁₆–2in)	Larval feeding habits

Order COLEOPTERA	Family COCCINELLIDAE	No. of species 5,000

LADYBIRDS

Also known as lady beetles or ladybugs, these
rounded, short-legged beetles are either shiny
and smooth or hairy. Most species are brightly
coloured black, red, yellow, or orange, often
with distinctive spots or stripes.
• LIFE-CYCLE Eggs are laid singly or in
small groups and glued to plants. The adults
and larvae of most species eat soft-bodied
insects. The larvae are dark in colour, and
the pupae may resemble bird droppings.
• OCCURRENCE Worldwide. On foliage.
• REMARK Ladybirds are beneficial control
agents of pest insects and mites.

three-segmented tarsi

rounded hind corner of pronotum

head hidden from view by pronotum

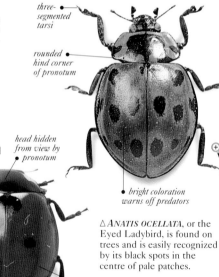

LARVAE are
often warty
or spiny.

bright coloration warns off predators

△ANATIS OCELLATA, or the
Eyed Ladybird, is found on
trees and is easily recognized
by its black spots in the
centre of pale patches.

COCCINELLA SEPTEMPUNCTATA
is a common, seven-spotted
European species that can
ooze liquid from its leg joints
to deter predators.

rounded, convex body

Length 0.1–1.5cm (¹⁄₃₂–⅝in)	Larval feeding habits

Order COLEOPTERA	Family CURCULIONIDAE	No. of species 48,000

WEEVILS

These insects are also called snout beetles, because of the extension of the head – the rostrum – that is present in most species, and which carries the mandibles. Many species have camouflage colouring, but others are brightly coloured with metallic scales.

• LIFE-CYCLE Females lay their eggs inside plant tissues. Many use their rostrums to bore holes in which to deposit the eggs.

• OCCURRENCE Worldwide. On most terrestrial and some aquatic plant species.

• REMARK Many weevils are pests of crops such as cotton and rice, and of bark. Weevils form the largest family in the animal kingdom.

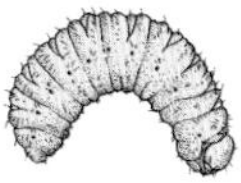

LARVAE are pale and grub-like, with a dark, tough head and no thoracic legs.

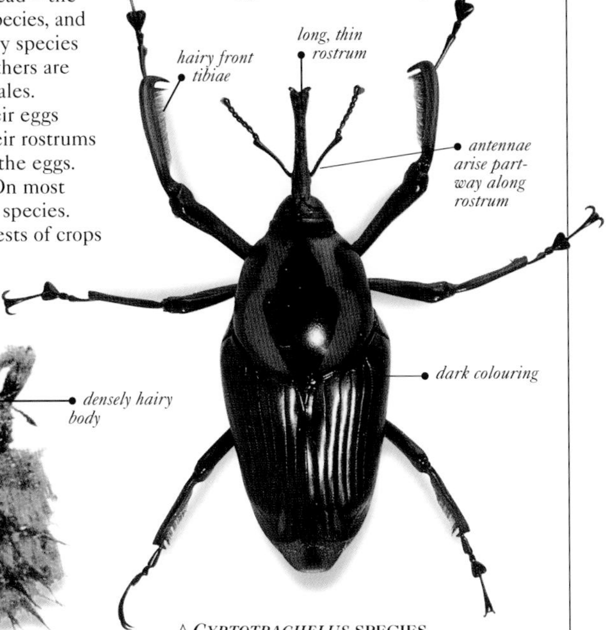

long, thin rostrum

hairy front tibiae

antennae arise part-way along rostrum

dark colouring

square-shouldered elytra

densely hairy body

▷ BRACHYCERUS FASCICULARIS is a strange-looking, hairy species that comes from South Africa. Its larvae feed on lilies.

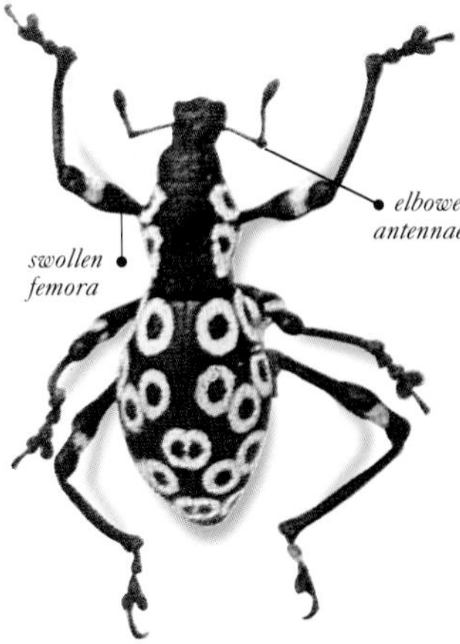

△ CYRTOTRACHELUS SPECIES come from Southeast Asia. The males have long, hairy front tibiae, used in courtship displays.

elbowed antennae

swollen femora

broad rostrum

shiny green coloration

expanded tarsal segments improve grip on smooth foliage

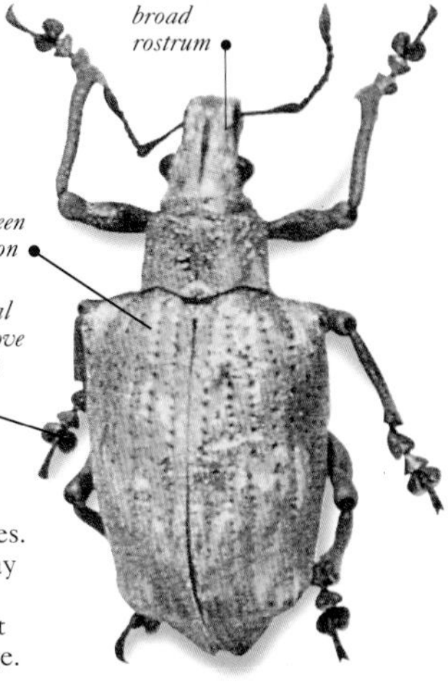

▷ LAMPROCYPHUS AUGUSTUS is a South American species. Its green colouring may serve to hide it from predators by helping it to blend in with foliage.

PACHYRHYNCHUS SPECIES have distinctive patterns of spots across their bodies. They are endemic to the Philippines.

Length 0.1–9cm (½–3½in)	Larval feeding habits

| Order COLEOPTERA | Family DERMESTIDAE | No. of species 950 |

LARDER BEETLES

Sometimes called skin or museum beetles, members of this family are rounded or slightly elongate. They are dull-coloured, brown or black, and are usually covered with patterns of coloured hairs or scales. The head may be difficult to see from above as it is often hidden by the pronotum.
• LIFE-CYCLE Eggs are usually laid on suitable foodstuff. The hairy, scavenging larvae feed at a rapid rate on a wide variety of organic matter.
• OCCURRENCE Worldwide. In a wide range of habitats.
• REMARK Many species are serious pests of foods stored in buildings, such as dried meat and spices, as well as textiles.

LARVAE are very hairy and are often referred to as "woolly bears".

mottled pattern of differently coloured scales

head hidden by pronotum when seen from above

△ *ANTHRENUS MUSEORUM*, the Museum Beetle, is so-called because its larvae can be pests of museum insect or plant collections.

DERMESTES LARDARIUS, the Larder Beetle, frequents food stores. The front half of its wing cases have white hairs and dark patches.

| Length 0.2–1.2cm (¹⁄₁₆–½in) | Larval feeding habits |

| Order COLEOPTERA | Family DYTISCIDAE | No. of species 3,500 |

PREDACIOUS DIVING BEETLES

These beetles are oval and streamlined, with smooth, shiny bodies. Most species are black or dark brown but some have yellow, brown, or green markings. The hindlegs are flat and fringed with hairs to provide propulsion. Highly adapted for aquatic life, these beetles can carry air supplies beneath their wing cases.
• LIFE-CYCLE The females cut slits in the stems of water plants and lay their eggs inside. Pupation occurs in wet soil near water. Like their parents, the larvae are fierce predators, attacking various prey, from insects to frogs and small fish.
• OCCURRENCE Worldwide. In streams, shallow lakes and ponds, brackish pools, and thermal springs.

LARVAE are elongate, with well-developed legs and large, curved mandibles.

• hairy hindlegs adapted for propulsion

• smooth, shiny, oval body is dark brown to black

DYTISCUS MARGINALIS, the Giant Diving Beetle, is found in Europe and Asia. The male has distinctive suction pads on its front legs for holding on to females during mating. Females have many more grooves along the elytra.

pronotum has yellow border

yellow legs

distinctive suction pads on front tarsi of males

| Length 0.2–4cm (¹⁄₁₆–1½in) | Larval feeding habits |

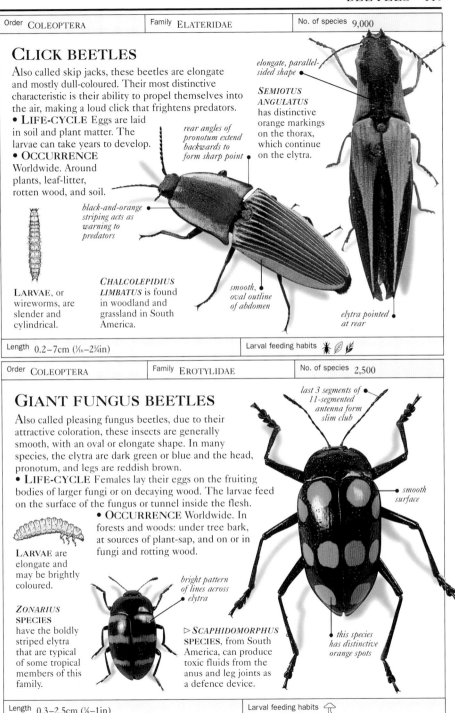

| Order COLEOPTERA | Family ELATERIDAE | No. of species 9,000 |

CLICK BEETLES

Also called skip jacks, these beetles are elongate
and mostly dull-coloured. Their most distinctive
characteristic is their ability to propel themselves into
the air, making a loud click that frightens predators.
• **LIFE-CYCLE** Eggs are laid
in soil and plant matter. The
larvae can take years to develop.
• **OCCURRENCE**
Worldwide. Around
plants, leaf-litter,
rotten wood, and soil.

*elongate, parallel-
sided shape*

*SEMIOTUS
ANGULATUS*
has distinctive
orange markings
on the thorax,
which continue
on the elytra.

*rear angles of
pronotum extend
backwards to
form sharp point*

*black-and-orange
striping acts as
warning to
predators*

LARVAE, or
wireworms, are
slender and
cylindrical.

*CHALCOLEPIDIUS
LIMBATUS* is found
in woodland and
grassland in South
America.

*smooth,
oval outline
of abdomen*

*elytra pointed
at rear*

| Length 0.2–7cm (¹⁄₁₆–2¾in) | Larval feeding habits 🌸 🍃 🪵 |

| Order COLEOPTERA | Family EROTYLIDAE | No. of species 2,500 |

GIANT FUNGUS BEETLES

*last 3 segments of
11-segmented
antenna form
slim club*

Also called pleasing fungus beetles, due to their
attractive coloration, these insects are generally
smooth, with an oval or elongate shape. In many
species, the elytra are dark green or blue and the head,
pronotum, and legs are reddish brown.
• **LIFE-CYCLE** Females lay their eggs on the fruiting
bodies of larger fungi or on decaying wood. The larvae feed
on the surface of the fungus or tunnel inside the flesh.
• **OCCURRENCE** Worldwide. In
forests and woods: under tree bark,
at sources of plant-sap, and on or in
fungi and rotting wood.

*smooth
surface*

LARVAE are
elongate and
may be brightly
coloured.

ZONARIUS
SPECIES
have the boldly
striped elytra
that are typical
of some tropical
members of this
family.

*bright pattern
of lines across
elytra*

▷ *SCAPHIDOMORPHUS*
SPECIES, from South
America, can produce
toxic fluids from the
anus and leg joints as
a defence device.

*this species
has distinctive
orange spots*

| Length 0.3–2.5cm (⅛–1in) | Larval feeding habits 🍄 |

Order COLEOPTERA	Family GEOTRUPIDAE	No. of species 600

DOR BEETLES

Also called earth-boring dung beetles, these stout insects are oval and rounded. They are brown or black and shiny, often with a metallic-blue or purple sheen. The tough elytra have longitudinal grooves or striations, and the broad tibiae of the front legs have strong "teeth" for digging.
• LIFE-CYCLE Eggs are laid in soil and dung. Adults prepare burrows and carry down dung, carrion, or decaying matter to feed their larvae.
• OCCURRENCE Worldwide. In dung, carrion, decaying wood, or fungi.

12-segmented antennae

last three segments form a club

tibiae of front legs have strong teeth for digging

LARVAE are grub-like and white and may be swollen towards the end of the abdomen.

horns on thorax of male

longitudinal grooves on elytra

shiny, tough elytra

rounded back

hind tibiae especially curved and toothed

TYPHAEUS TYPHOEUS is known as the Minotaur Beetle because the males, shown here, have horns. Females make burrows, stocking the side-tunnels with rabbit dung.

▷ **CERATOPHYUS HOFFMANNSEGGI** comes from Spain, Portugal, and North Africa. It buries the dung of various herbivores.

Length 0.4–4cm (⁵⁄₃₂–1½in)	Larval feeding habits

Order COLEOPTERA	Family GYRINIDAE	No. of species 750

WHIRLIGIG BEETLES

These oval, streamlined beetles are mostly black, often with a bronze or steel-blue sheen. The long front legs are used to grasp prey, while the middle and hindlegs are short and paddle-like. Whirligigs occur in large numbers on the surface of water and use their ripple-sensitive antennae to locate prey.
• LIFE-CYCLE Groups of eggs are laid on the underside of leaves. Pupation occurs on land, in a cocoon inside a mud cell.
• OCCURRENCE Worldwide. On the surface of ponds and slow-flowing streams.
• REMARK Their name derives from the fact that they swim rapidly, in circles.

eyes divided into upper and lower parts

short antennae

long front legs used to grasp prey

short, paddle-like middle and hindlegs

LARVAE are elongate, with sharp, sucking mandibles and feathery abdominal gills.

ENHYDRUS SPECIES are found in South America. Like all whirligig beetles, they produce a fruity odour that deters predators.

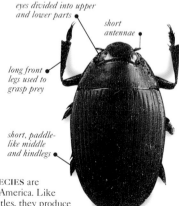

Length 0.3–2.5cm (⅛–1in)	Larval feeding habits

Order COLEOPTERA	Family HISTERIDAE	No. of species 3,000

HISTER BEETLES

These tough-bodied insects are rounded or oval. Many
are convex in profile, although some have flat bodies that
allow them to live under bark. Most species are black
and have various striations and punctures on the
surface of the elytra and other body parts.

• LIFE-CYCLE Eggs are laid in carrion,
dung, and rotting plant matter. The larvae
(and adults) eat other insects, especially
fly maggots and beetle grubs.

• OCCURRENCE Worldwide. In
dung, carrion, or leaf-litter; under bark; in
tunnels of wood-boring insects; in nests of
ants or birds; or inside termite colonies.

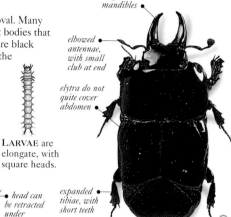

*large, curved
mandibles*

*elbowed
antennae,
with small
club at end*

*elytra do not
quite cover
abdomen*

LARVAE are
elongate, with
square heads.

*very shiny
surface*

SIDE VIEW

*head can
be retracted
under
pronotum*

*expanded
tibiae, with
short teeth*

CARCINOPS PUCILO is
a black, extremely shiny
beetle that breeds in
rotting vegetable or animal
matter and in bat dung.

*longitudinal
grooves on surface
of elytra*

OVERHEAD

△ **HOLOLEPTA** SPECIES are
very flat, with an extremely
hard body surface. They hunt
for prey underneath the bark
of various trees.

Length 0.1–2.5cm (½–1in)	Larval feeding habits

Order COLEOPTERA	Family HYDROPHILIDAE	No. of species 2,000

WATER SCAVENGER BEETLES

These beetles are oval and coloured black, brown, or
sometimes yellowish. The principal distinguishing
feature is their antennae, which end in a four-
segmented club. The last three segments of this
club are hairy, whereas the first is smooth and
saucer-shaped. Many adults live in water, carrying air
under their wing cases and on the surface of their body.

• LIFE-CYCLE Eggs are laid on water and in dung
and damp places. Adults are usually scavengers; the
larvae eat aquatic insect larvae, snails, and worms.
The larvae may have gills or take air at the surface.

• OCCURRENCE Worldwide. Most live in water
and damp habitats, although some are found
in dung, decaying vegetation, and soil.

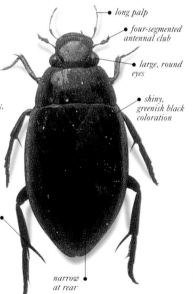

long palp

*four-segmented
antennal club*

*large, round
eyes*

*shiny,
greenish black
coloration*

*large spine
on tibia*

HYDROPHILUS PICEUS, also
called the Great Silver Water
Beetle, uses its specially shaped
antennae to channel air back,
under the wings.

*narrow
at rear*

LARVAE may
have warty or
hairy backs.

Length 0.1–4.5cm (½–1¾in)	Larval feeding habits

Order COLEOPTERA	Family LAMPYRIDAE	No. of species 2,000

FIREFLIES

Also called lightning bugs, fireflies are neither flies nor bugs but flat, elongate, or slightly oval beetles. Most are dull-coloured but may have red or yellow markings. Males usually have fully developed wings, whereas females may be wingless. Some wingless females look like larvae. The common names derive from the fact that the adults of many species communicate with mates by using species-specific flashes of cold, green light. These are made by luminous organs on the underside of the abdomen.

• LIFE-CYCLE Eggs are laid on vegetation. The larvae, which are commonly known as glow worms, feed on invertebrates and snails.

• OCCURRENCE Worldwide. On vegetation in woodland and moist grassland.

• REMARK Females of some species imitate the flashing of closely related species, luring the males with their sexual signals and then eating them.

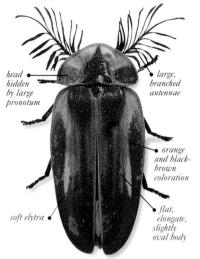

head hidden by large pronotum

large, branched antennae

orange and black-brown coloration

soft elytra

flat, elongate, slightly oval body

LARVAE are elongate and taper at both ends. The head is small and usually longer than its width.

LAMPROCERA SELAS has distinctive antennae and orange-red and black-brown coloration, which warns potential predators that this insect is distasteful.

Length 0.5–3cm (³⁄₁₆–1¼in)	Larval feeding habits

Order COLEOPTERA	Family LATHRIDIIDAE	No. of species 500

MINUTE SCAVENGER BEETLES

These tiny, oval, brown or black beetles, also called mould beetles, have a small, rounded pronotum. The wing cases have rows of ribs or puncture marks, and may be slightly hairy or bristly.

• LIFE-CYCLE Eggs are laid on substances such as decaying matter and fungi. The larvae feed on the spores of various fungi.

• OCCURRENCE Worldwide, especially in temperate regions. Under stones and bark; in fungi, decaying or mouldy material, and birds' nests; and on flowers.

antennae have 11 segments and small, 2–3 segmented club

◁ARIDIUS BIFASCIATUS is a fungus-feeding native of Australia, now found elsewhere. Its elytra have distinctive coarse grooves and dark markings.

small, rounded pronotum

oval outline

longitudinal rows of puncture marks on elytra

LARVAE are pale, oval or elongate, and slightly flat. The upper surface has groups of hairs.

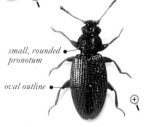

ENICMUS TRANSVERSUS has brown to black coloration, a very flat body, and distinctive grooves on its elytra.

CORTICARIA IMPRESSA is a dark brown, fungus-feeding beetle. Its wing cases are a lighter colour than the head and thorax.

Length 1–3mm (¹⁄₃₂–¹⁄₈in)	Larval feeding habits

Order COLEOPTERA	Family LUCANIDAE	No. of species 1,300

STAG BEETLES

These beetles are typically large, smooth, and black or reddish brown. Males are larger than females and have greatly enlarged mandibles that may be lined with prominent teeth. These powerful mandibles are used in male-to-male combat over females and are designed to fit around the edges of a rival's pronotum, with the victor flipping its rival over. Adults are mainly nocturnal and either do not feed or take fluids (plant sap, nectar, or fruit juice).
• LIFE-CYCLE The female lays her eggs on decaying tree stumps, roots, or logs. Stag beetle larvae may take several years to complete their development. The larvae pupate inside a cell that is composed of chewed wood fibres.
• OCCURRENCE Worldwide. In deciduous woodland and forests.

large mandibles bent at right angles •

• large head

eyes at side • of head

elbowed • antennae with three- or four-segmented club

▷ *MESOTOPUS TARANDUS* is found in Africa. Its large, toothed mandibles are bent almost at right angles, to lock around the bodies of rival males when fighting.

• large, shiny black body

LARVAE are C-shaped, with strong thoracic legs.

• male has very long, straight, toothed mandibles

head smaller • than pronotum

strong front leg with spined tibia

• sharp spines on rear corners of pronotum

shiny, metallic elytra •

CHIASOGNATHUS GRANTI males have greatly elongated mandibles, which they use in male-to-male combat.

PHALACROGNATHUS MULLERI is a distinctively coloured, metallic stag beetle that is found in the rainforests of northern Australia.

Length 0.5–8cm (³⁄₁₆–3¼in), most 1.5–4.5cm (⅝–1¾in)	Larval feeding habits

Order COLEOPTERA	Family LYCIDAE	No. of species 3,500

NET-WINGED BEETLES

The wings of these soft-bodied, black-and-red or yellow beetles often have a net-like pattern of cells on the elytra. Adult females of a few rainforest species look like very large larvae.
• **LIFE-CYCLE** Eggs are typically laid in or on the larval feeding site – the larvae suck liquids from rotting matter or eat small arthropods. There is some debate about what these beetles eat, and little is known about the life-cycle of most species. Adults probably do not feed a great deal, but some take nectar and pollen.
• **OCCURRENCE** Worldwide, except in New Zealand, mainly in warmer regions. In wooded and well-vegetated areas.

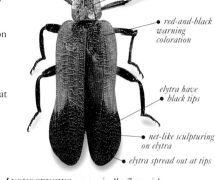

head hidden from above by pronotum

red-and-black warning coloration

elytra have black tips

net-like sculpturing on elytra

elytra spread out at tips

LARVAE are often wider in the middle than at the ends, and the dorsal surface may be patterned.

LYCUS SPECIES are typically flat, with black-and-red warning coloration that deters predators. Males and females may have different shapes or markings.

Length 0.3–3cm (⅛–1¼in)	Larval feeding habits 🐜 🌿

Order COLEOPTERA	Family MELOIDAE	No. of species 3,000

BLISTER BEETLES

Also called oil beetles, the adults produce cantharidin, an oily fluid to deter predators that can blister human skin. Most species are soft, leathery, and black or brown with red or yellow marks. Some are metallic. Most are long and parallel-sided; a few are oval. The elytra vary in length.
• **LIFE-CYCLE** Eggs are laid in soil. The mobile, first-stage larvae locate and eat the eggs of grasshoppers or bees. Adults are herbivorous.
• **OCCURRENCE** Worldwide, except in New Zealand, mainly in warm, dry areas. On flowers and foliage.
• **REMARK** Some species are crop pests. Cantharidin from *Lytta vesicatoria* is used to treat urogenital disorders.

square or narrow pronotum

head points downwards

△ *MYLABRIS* SPECIES are all brightly coloured and secrete toxic or blistering fluids from their leg joints. Some species are serious pests of millet crops in West Africa.

LYTTA VESICATORIA, the Spanish Fly, is a bright, iridescent green and produces a mouse-like odour. Its larvae develop in the nests of solitary bees.

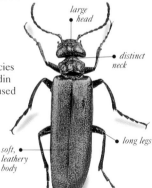

large head

distinct neck

soft, leathery body

long legs

LARVAE become increasingly grub-like with each moult.

Length 0.5–3cm (¾₆–1¼in), most 1–2cm (⅜–¾in)	Larval feeding habits 🐜 🕷

Order COLEOPTERA	Family MORDELLIDAE	No. of species 1,500

TUMBLING FLOWER BEETLES

These beetles, named after their habit of tumbling when disturbed, are typically humped or arched in profile but quite flat from side to side. The end of the abdomen is spine-like. Most species are brown or black with patterns of short white, red, or yellowish hairs or scales.

• **LIFE-CYCLE** Eggs are laid singly in dead wood or plants. The larvae develop mostly in decaying wood or mine plant stems. Adults eat nectar.

• **OCCURRENCE** Worldwide, mainly in tropical regions. Adults gather on flowers and dead wood in the sun.

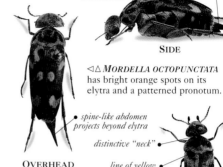

stout thorax

tapering abdomen

SIDE

◁△ *MORDELLA OCTOPUNCTATA* has bright orange spots on its elytra and a patterned pronotum.

spine-like abdomen projects beyond elytra

distinctive "neck"

line of yellow hairs at rear edge of thorax

two irregular yellow bands on elytra

OVERHEAD

LARVAE are cylindrical and elongate, with a round head and short legs.

▷ *TOMOXIA BUCEPHALA* has the parallel-sided to slightly tapering shape that is characteristic of this family.

Length 0.2–1.5cm (1⁄16–5⁄8in)	Larval feeding habits

Order COLEOPTERA	Family NITIDULIDAE	No. of species 3,000

POLLEN BEETLES

Also known as sap beetles, these small beetles are mostly oval, square, or rectangular, and are often convex in profile. Most are brown or black and can have irregular red or yellow spots. In some species the elytra are shortened, exposing the last couple of abdominal segments.

• **LIFE-CYCLE** Eggs are laid around feeding sites – adults and larvae feed on sap, nectar, and pollen, but a few prey on scale insects, and some larvae develop inside seed pods.

• **OCCURRENCE** Worldwide. At flowers, sap flows, and decaying plant and animal material; a few species live in ant and bee colonies.

LARVAE are long, slender, slightly curved, and white.

EPURAEA **SPECIES** are mostly pale or red-brown and slightly flat, with clubbed antennae and dark eyes.

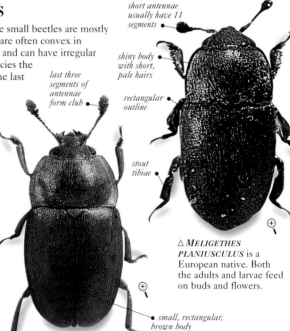

short antennae usually have 11 segments

shiny body with short, pale hairs

rectangular outline

last three segments of antennae form club

stout tibiae

△ *MELIGETHES PLANIUSCULUS* is a European native. Both the adults and larvae feed on buds and flowers.

small, rectangular, brown body

Length 0.1–1.5cm (1⁄32–5⁄8in)	Larval feeding habits

Order COLEOPTERA	Family PASSALIDAE	No. of species 500

BESS BEETLES

Also called betsy or patent-leather beetles,
these species are shiny black or dark brown with
a parallel-sided, flattened body. The elytra have
noticeable striations, and many species rarely fly.
• **LIFE-CYCLE** Eggs are laid in dead wood. Adults
and larvae live together in rotten tree stumps and
logs. Adults chew wood into pulp for the larvae to
eat – the cellulose is digested by fungi in the gut.
• **OCCURRENCE** Primarily tropical regions. Mainly
in wooded areas of South America and Southeast
Asia. Some are found inside leaf-cutter ant nests.
• **REMARK** Adults make sounds by rubbing their
hindwings against the abdomen. These are used for
aggression, mating, and keeping in touch with larvae.

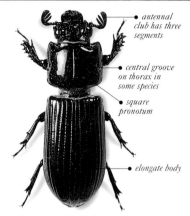

*antennal
club has three
segments*

*central groove
on thorax in
some species*

*square
pronotum*

elongate body

AULACOCYCLUS PARRYI has a
flattened body that allows it to dig
through detritus and under bark.

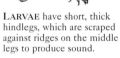

LARVAE have short, thick
hindlegs, which are scraped
against ridges on the middle
legs to produce sound.

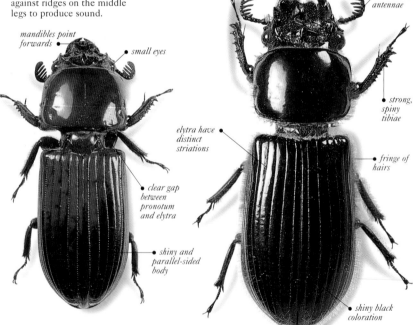

*mandibles point
forwards*

small eyes

*comb-like
antennae*

*strong,
spiny
tibiae*

*elytra have
distinct
striations*

*fringe of
hairs*

*clear gap
between
pronotum
and elytra*

*shiny and
parallel-sided
body*

*shiny black
coloration*

AULACOCYCLUS **SPECIES** are
well represented across Southeast
Asia. Individual species look very
similar, and detailed examination is
required for correct identification.

ACERAIUS RECTIDENS has
curled, ten-segmented antennae
and a short peg on the head.

Length 1–8.5cm (⅜–3¼in)	Larval feeding habits

Order COLEOPTERA	Family PSELAPHIDAE	No. of species 9,000

PSELAPHID BEETLES

The abdomen of these small brown beetles is considerably broader than the pronotum and the head. The elytra are short, which leaves most of the abdomen exposed.

• **LIFE-CYCLE** Little is known about the life-cycle of most species. Adults, and presumably larvae, eat mites, small insects, and other invertebrates.

• **OCCURRENCE** Worldwide. Under bark, in rotting vegetation and wood, and in moss, leaf-litter, caves, and soil. Some species live in ant colonies.

▷ *RYBAXIS LONGICORNIS* is European and lives in moss clumps near pools and bogs.

short elytra leave abdomen exposed

last segment of antennae enlarged

◁ *BATRISODES DELAPORTI* lives in association with *Lasius brunneus* ants, in dead wood.

LARVAE are often slightly flattened in shape.

Length 0.5–6mm (¹⁄₆₄–¹⁄₄in), most under 3mm (¹⁄₈in)	Larval feeding habits 🐾

Order COLEOPTERA	Family PYROCHROIDAE	No. of species 150

PYROCHROID BEETLES

Also called fire-coloured beetles, because of their distinctive black, red, and yellow colouring, most species are narrow and soft-bodied. The eyes are large in relation to the head, and the segments of the antennae are serrated or have long processes.

• **LIFE-CYCLE** Eggs are laid under the bark of dead wood, where the larvae develop, eating threads of fungi or small creatures.

• **OCCURRENCE** Mostly in northern hemisphere and Southeast Asia. In woodland and forests.

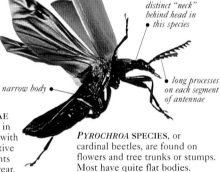

distinct "neck" behind head in this species

narrow body

long processes on each segment of antennae

LARVAE are flat in shape, with distinctive segments at the rear.

PYROCHROA **SPECIES**, or cardinal beetles, are found on flowers and tree trunks or stumps. Most have quite flat bodies.

Length 0.6–1.8cm (¼–¾in)	Larval feeding habits 🐾 🍄

Order COLEOPTERA	Family RHIPIPHORIDAE	No. of species 450

WEDGE-SHAPED BEETLES

These hump-backed, black-and-orangish beetles are described as wedge-shaped because of their extremely blunt-ended abdomens. The wings may be either full-sized or short.

• **LIFE-CYCLE** Eggs are laid where hosts visit – the larvae are parasitic on cockroaches or the larvae of wood-boring beetles, bees, and wasps. Young larvae may have legs. Later stages are maggot-like.

• **OCCURRENCE** Worldwide. In a variety of habitats where insect hosts can be found.

METOECUS PARADOXUS is the only British species in this family. Its larvae are parasites of *Vespula* wasps and their larvae.

LARVAE of early stages may have legs to grip the host insect.

wedge-shaped body

Length 0.4–3.6cm (⁵⁄₃₂–1⅖in)	Larval feeding habits 🐾

Order COLEOPTERA	Family SCARABAEIDAE	No. of species 16,500

SCARABS

The huge family Scarabaeidae comprises several major subfamilies, including dung beetles or scarabs, giant hercules beetles, and leaf chafers. Extremely varied in shape and size, their colours range from black to metallic blues and greens. All species have antennae that end in a kind of club, made up of several movable plates. The males of many species have horns, which they use when fighting over females.

• **LIFE-CYCLE** Eggs are laid – and larvae can be found – in soil, the dung of herbivorous mammals, rotten wood, and decaying matter. Some adult dung beetles use their hindlegs to roll fresh dung away in balls before burying the balls and laying their eggs inside. Others bury dung where it is.

• **OCCURRENCE** Worldwide. On dung, carrion, decaying matter, fungi, and vegetation; under bark; and in the burrows of vertebrates or in ant and termite nests.

• **REMARK** The dung or scarab beetle was a major religious symbol in Ancient Egypt. Mummification may have been an imitation of its pupal stage, protecting the body during its transformation and ultimate rebirth.

bright green, shiny, metallic colouring

△ *PHANEUS DEMON* is a Central American species whose larvae develop inside the dung of large herbivores.

shiny, golden colouring

strong claws

last abdominal segments showing

stout hind femora

▷ *PLUSIOTIS RESPLENDENS* is found in Central America. Many brightly coloured species such as this are much sought after by unscrupulous collectors.

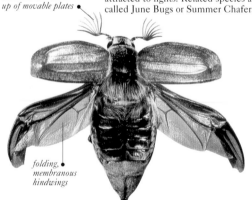

▽ *MELOLONTHA MELOLONTHA*, the Cockchafer, is nocturnal and often attracted to lights. Related species are called June Bugs or Summer Chafers.

antennal club made up of movable plates

folding, membranous hindwings

shovel-shaped head

red-purple, metallic sheen

△ *KHEPER AEGYPTIORUM*, found in Africa, rolls dung balls as large as 5cm (2in) away from dung pats to bury. This large, colourful beetle was probably the first to be revered in Ancient Egypt.

Length 0.2–17cm (1/16–6¾in)	Larval feeding habits

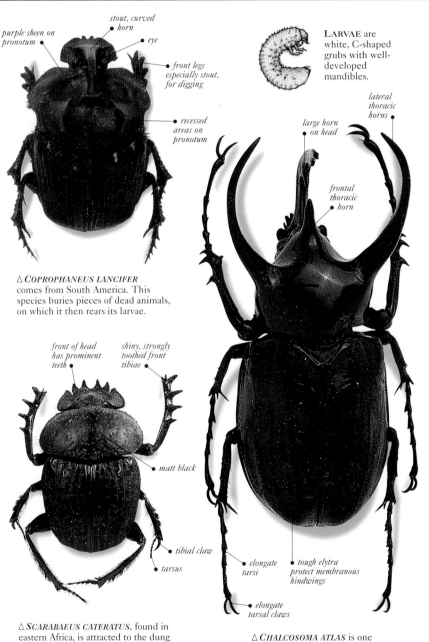

purple sheen on pronotum

stout, curved horn

eye

front legs especially stout, for digging

recessed areas on pronotum

LARVAE are white, C-shaped grubs with well-developed mandibles.

large horn on head

lateral thoracic horns

frontal thoracic horn

△ *COPROPHANEUS LANCIFER* comes from South America. This species buries pieces of dead animals, on which it then rears its larvae.

front of head has prominent teeth

shiny, strongly toothed front tibiae

matt black

tibial claw

tarsus

elongate tarsi

tough elytra protect membranous hindwings

elongate tarsal claws

△ *SCARABAEUS CATERATUS*, found in eastern Africa, is attracted to the dung of large herbivores, such as buffalo and giraffes. Females stay in the nest to look after the young when they emerge from within the brooding dung balls.

△ *CHALCOSOMA ATLAS* is one of three known species from eastern Asia. Males use their horns to grapple with each other in contests over females.

Order COLEOPTERA	Family SILPHIDAE	No. of species 250

CARRION BEETLES

Many of these flat, soft-bodied species are black or brown, often with bright yellow, red, or orange markings. In some species, the elytra are shortened, exposing several abdominal segments.
• **LIFE-CYCLE** Adults and larvae are mostly scavengers, eating rotting animal or plant material. Species of the genus *Nicrophorus* (sexton or burying beetles) bury corpses of small animals and lay their eggs on the buried carcass. In some species, adults may feed their larvae regurgitated carrion.
• **OCCURRENCE** Worldwide, but mainly in the northern hemisphere. On the ground near carcasses, dung, and rotting fungi.

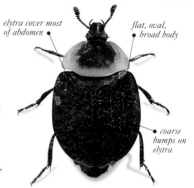

elytra cover most of abdomen

flat, oval, broad body

coarse bumps on elytra

LARVAE are flat and elongate, with a broad pronotum and small head.

SILPHA AMERICANA is a native of North America. It has a broad shape and fairly bright coloration. Like all carrion beetles, it is quickly attracted to carrion by the odour.

Length 0.4–4.5cm (⁵⁄₃₂–1¾in), most under 2cm (¾in)	Larval feeding habits

Order COLEOPTERA	Family STAPHYLINIDAE	No. of species 29,000

ROVE BEETLES

Most of these beetles are small and smooth, with elongate, parallel-sided, brown or black bodies. Some may have a sculptured body surface, bright colours, or body hairs. They all have short elytra, and a highly mobile, exposed abdomen. Small species tend to be diurnal; larger species are usually nocturnal.
• **LIFE-CYCLE** Eggs are commonly laid in soil, fungi, and leaf-litter. Most larvae prey on insects and other arthropods, and usually live in the same place as the adults.
• **OCCURRENCE** Worldwide. In soil, fungi, leaf-litter, decaying plants, and carrion. Some are found in ant or termite colonies or in the fur of some mammals.
• **REMARK** Species of the genus *Paederus* can blister the skin if handled.

matt black all over

flexible abdomen

flat head and flat first segment of thorax

STAPHYLINUS OLENS, or the Devil's Coach Horse, is a large, black, European species. If disturbed or threatened, it curves its abdomen up in a scorpion-like threat posture.

abdomen used to push hindwings under elytra when wings not in use

covering of black and yellow hairs

hindwings folded beneath short elytra

elongate mandibles

LARVAE are elongate, with short antennae and cerci.

EMUS HIRTUS is a distinctive, large species, native to southern Europe. It likes to hunt prey that feeds on carrion or dung.

Length 0.1–4cm (¹⁄₂–1½in), most under 2cm (¾in)	Larval feeding habits

Order COLEOPTERA	Family TENEBRIONIDAE	No. of species 17,000

DARKLING BEETLES

These beetles are mostly black or brown in colour. Some species, however, have coloured markings or white elytra. There is a great variation in shape within this family, from parallel-sided and blunt-ended to large and broadly oval. The body may be smooth and shiny or dull and roughly textured.

• LIFE-CYCLE The eggs are scattered singly or in groups in and around the larval feeding-matter. These beetles are scavengers and mostly eat decaying vegetable or animal material; some larvae will eat plant roots. Adults of many species are able to produce a foul-smelling secretion, used for defence, from special glands on the abdomen.

• OCCURRENCE Worldwide. In all terrestrial habitats, especially desert and arid regions.

• REMARK Some species can be pests. The family includes flour beetles – which may damage stored, dried foods such as flour, grain, and cereals – while other species damage coffee or mushroom crops.

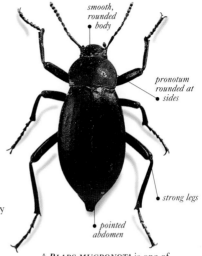

smooth, rounded body

pronotum rounded at sides

strong legs

pointed abdomen

△ *BLAPS MUCRONOTA* is one of six similar, non-flying European species. Called cellar or churchyard beetles, they favour dark places.

LARVAE are elongate and cylindrical with tough bodies and short legs.

antennae usually have 11 segments

white elytra reflect the sun's heat

long legs for running over hot sand

ONYMACRIS CANDIDIPENNIS has elytra with no pigmentation. Combined with its long legs, this allows it to be active during the day in its native Namib Desert.

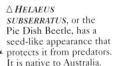

thread-like antennae

flat expansions of elytra and pronotum

oval body

△ *HELAEUS SUBSERRATUS*, or the Pie Dish Beetle, has a seed-like appearance that protects it from predators. It is native to Australia.

elytra of this newly emerged adult not yet developed

short, strong legs

brown coloration

TENEBRIO MOLITOR, or the Yellow Mealworm Beetle, is found across the world in stored grains and flour.

Length 0.2–5cm (¹⁄₁₆–2in), most under 2cm (¾in)	Larval feeding habits

STREPSIPTERANS

T HERE ARE 8 FAMILIES and 560 species in the order Strepsiptera. Male strepsipterans have large, fan-shaped hindwings and tiny, strap-like forewings. The hindwings have a twisted appearance, giving rise to the other common name for this order – twisted-winged parasites. The grub-like females are wingless and, typically, legless. They usually live as endoparasites, inside the bodies of other insects. Strepsipterans use species from many different orders as hosts but favour bugs, wasps, and bees.

Males detect mates by the sexual pheromones that females emit and then cling to the host's body and mate with the female inside. Metamorphosis is complete. Eggs hatch inside the female. Hundreds or thousands of tiny, six-legged larvae (triungulins) emerge from the host, and each goes in search of its own host. Once a host is found, the larva uses an enzyme to get inside and moults to become a legless endoparasite, feeding on the host's body fluids and tissues. Pupation occurs inside the host. Males emerge and fly off, killing the host in the process, but females remain inside the live host.

Some species of strepsipterans are useful control agents of pests such as crickets and planthoppers.

Order STREPSIPTERA	Family STYLOPIDAE	No. of species 260

STYLOPIDS

Males of this family are small and dark, with bulging, berry-like eyes. The forewings are tiny and strap-like; the hindwings large and fan-like. Females are wingless and legless and never leave the body of their host.

• LIFE-CYCLE Most stylopid species parasitize mining or sweat bees or solitary or vespid wasps (see pp.178–93). After mating, eggs hatch inside the female stylopid. The active triungulin larvae then leave the female through a special brood passage and crawl on to flowers to wait for the next host.

• OCCURRENCE Worldwide. In various well-vegetated habitats, wherever hosts are found. Males are free-living, whereas the females of the species live in the bodies of certain bees and wasps.

• REMARK The sexual organs of the parasitized hosts degenerate and, in some cases, there may be a reversal of secondary sexual characters, so that males look like females and vice versa.

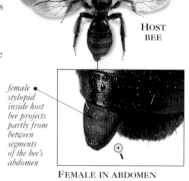

HOST BEE

female ● stylopid inside host bee projects partly from between segments of the bee's abdomen

FEMALE IN ABDOMEN OF HOST BEE

LARVAE are very small, with a pair of bristles at the end of the abdomen that they use to jump.

branched antennae ●

odd, berry-like eyes

MALE

STYLOPS SPECIES have a highly distinctive shape. They have an interesting biology that has led to them being used as the emblem of a major entomological society.

● hindwings have slightly twisted appearance

Length 0.5–4mm (¼₄–⁵⁄₃₂in)	Larval feeding habits

SCORPIONFLIES

THE ORDER MECOPTERA includes 9 families and 550 species. The common name refers to the scorpion-like abdomen seen in the males of certain species – slender and upturned, with swollen genitalia. Scorpionflies have an elongated body, and most species have two pairs of narrow wings. The head is typically lengthened downwards to form a beak, called the rostrum, which bears the jaws.

Scorpionflies feed on dead or dying insects and will also feed on carrion, nectar, or various other fluids. Mating often occurs after dark. It may involve the presentation of nuptial gifts by the males, usually in the form of a dead insect or a mass of saliva that the male produces. Females will reject males who offer small or poor gifts, but a male may then simply take a mate by force, seizing the female with its genital claspers. Sexual pheromones may also be involved in the courtship rituals, being produced either by the males only or by both sexes.

Scorpionflies lay their eggs in soil. Metamorphosis is complete and the larvae are either highly caterpillar-like, with abdominal prolegs, or grub-like. Pupation takes place in an underground cell or in vegetation.

Order MECOPTERA	Family BITTACIDAE	No. of species 170

HANGINGFLIES

These scorpionflies can look very like crane flies (see p.140). They are characterized by slender wings, very long legs, and specially modified hind tarsi for capturing prey. The fifth tarsal segment of the hindleg is sharp and can grip prey tightly. Hangingflies typically hang from vegetation by their front legs and trail their long hindlegs to catch passing prey.

• **LIFE-CYCLE** The mating process can be complicated by the males stealing each other's nuptial gifts. Eggs are laid in soil. The caterpillar-like larvae stick debris to their bodies for camouflage. They crawl around after dark on soil and leaf-litter, eating dead insects.

• **OCCURRENCE** Southern hemisphere. In damp woodland or well-vegetated, shady areas.

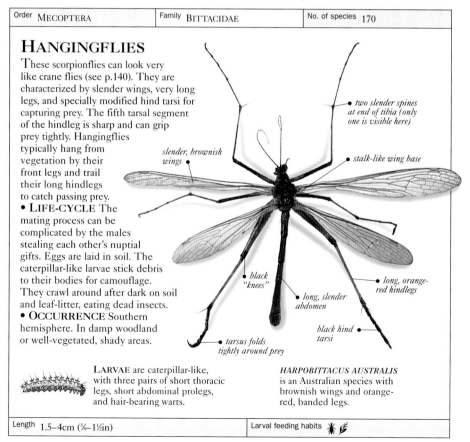

two slender spines at end of tibia (only one is visible here)

stalk-like wing base

slender, brownish wings

black "knees"

long, orange-red hindlegs

long, slender abdomen

black hind tarsi

tarsus folds tightly around prey

LARVAE are caterpillar-like, with three pairs of short thoracic legs, short abdominal prolegs, and hair-bearing warts.

HARPOBITTACUS AUSTRALIS is an Australian species with brownish wings and orange-red, banded legs.

Length 1.5–4cm (⅝–1½in)	Larval feeding habits

Order MECOPTERA	Family BOREIDAE	No. of species 26

SNOW SCORPIONFLIES

These dark brown or black insects do not fly, although their long middle and hindlegs allow them to jump short distances. The wings are reduced and hook-like in the males, and scale-like in the females.

• LIFE-CYCLE Eggs are laid in moss. The adults and larvae of this family mainly feed on mosses and lichens.

• OCCURRENCE Northern hemisphere, in cold, mountainous areas. They occur on snow, in mosses, and under stones, but are not very common.

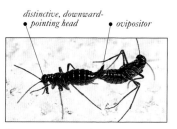

distinctive, downward-pointing head • ovipositor

LARVAE are cylindrical and caterpillar-like, with well-developed thoracic legs.

BOREUS BRUMALIS is seen here mating on the surface of snow. This is the commonest snow scorpionfly in northeastern parts of the USA.

Length 2–5mm (¹⁄₁₆–⁹⁄₃₂in), most 3–5mm (⅛–³⁄₁₆in)	Larval feeding habits

Order MECOPTERA	Family PANORPIDAE	No. of species 360

COMMON SCORPIONFLIES

The wings of these scorpionflies usually have brown or black patterning. In females, the abdomen tapers to a point. In males, the genital apparatus at the end of the abdomen is swollen and upturned.

• LIFE-CYCLE Some males steal prey from spiders' webs to feed the female during mating. Eggs are laid in soil or damp leaf-litter, and the larvae resemble caterpillars. Pupation takes place inside a cell underground. Adults eat dead insects, nectar, and fruit.

• OCCURRENCE Worldwide, but mostly in the northern hemisphere. Among vegetation in a variety of shady habitats.

broad white bands on • forewings

broad white • bands on hindwings

hindwings similar in shape to forewings, but slightly smaller

• slender antennae

slender • legs

PANORPA NUPTIALIS is a large, distinctive North American species with striking white bands across both pairs of wings.

genitalia •

• brown wings with white patches

◁ PANORPA LUGUBRIS is common in parts of North America. The males of this large species are bigger than the females.

LARVAE have three pairs of thoracic legs and eight pairs of short, abdominal prolegs.

• upturned, swollen genitals in male

Length 0.9–2.5cm (¹¹⁄₃₂–1in), most 1–1.5cm (⅜–⅝in)	Larval feeding habits

FLEAS

THE ORDER SIPHONAPTERA is divided into 18 families and 2,000 species of flea. These brown, shiny, and wingless insects have tough, laterally flattened bodies covered with backward-pointing spines and bristles. The enlarged hindlegs are part of a unique jumping mechanism involving energy storage in rubber-like pads of protein. Fleas have specialized mouthparts for sucking blood. They are ectoparasites, living on the outside of a host animal and feeding on it without killing it.

Less host-specific than parasitic lice (see pp.83–84), some fleas are found on over 30 host species – mostly terrestrial mammals but also birds. Metamorphosis is complete. The larvae do not suck blood, but scavenge on detritus, dried blood, and the excrement of adult fleas.

Order SIPHONAPTERA	Family HYSTRICHOPSYLLIDAE	No. of species 80

RODENT FLEAS

These species of fleas have a comb of stout, dark bristles at the rear edge of the pronotum. Most are ectoparasites of small rodents such as mice and wood rats.
• LIFE-CYCLE Sticky eggs are laid in the host's burrow, lair, or nest. There are three larval stages, and pupation takes place within a silken cocoon.
• OCCURRENCE Worldwide, mainly in the northern hemisphere. In hosts' nests.
• REMARK Pest species include *Ceratophyllus gallinae*, the European Chicken Flea.

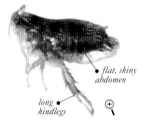

flat, shiny abdomen

long hindlegs

LARVAE are slender, pale, relatively hairless, and without legs.

EPITEDIA SPECIES are found only in North America. Their favoured hosts are rats, mice, and shrews.

Length 1–4mm (¹⁄₃₂–³⁄₃₂in)		Larval feeding habits

Order SIPHONAPTERA	Family PULICIDAE	No. of species 200

COMMON FLEAS

This family looks typical of its order and may have bristle-combs on its pronotum and cheeks. Common fleas are ectoparasitic on humans and a wide range of other mammals, including dogs, cats, and rabbits.
• LIFE-CYCLE Eggs are dropped in hosts' nests or burrows. Adults can survive for a long time without a blood meal. Emergent fleas remain in their cocoon until they sense a host's presence.
• OCCURRENCE Worldwide. On mammalian hosts in a wide range of habitats.
• REMARK Many species spread disease. The Dog Flea carries a tapeworm that affects dogs, cats, and humans. The bacterium that caused bubonic plague in medieval Europe was carried by various types of rat flea.

▷ *SPILOPSYLLUS CUNICULI*, the Rabbit Flea, is widespread in the northern hemisphere and carries the rabbit disease myxomatosis.

eye touches cheek
comb

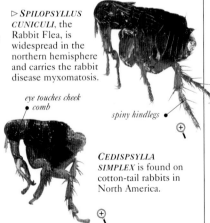

spiny hindlegs

CEDISPSYLLA SIMPLEX is found on cotton-tail rabbits in North America.

LARVAE are slender, pale, and worm-like.

Length 1–8mm (¹⁄₃₂–⁵⁄₁₆in)		Larval feeding habits

TWO-WINGED FLIES

THE ORDER DIPTERA contains 130 families 'and 122,000 species. These insects have just one pair of wings, as the hindwings are reduced to small, club-shaped balancing organs called halteres. Some are wingless. There are two sub-orders. The delicate and slender Nematocera (Bibionidae to Tipulidae below) have slim antennae. The Brachycera, divided into Orthorrhapha and Cyclorrhapha (and represented here by the families Acroceridae to Tephritidae) are more robust, with short, stout antennae that have fewer than six segments. Metamorphosis is complete.

Two-winged flies are vital pollinators, parasites, predators, and decomposers in all kinds of habitats. Many, however, damage crops or carry diseases that have a huge impact on animals and humans.

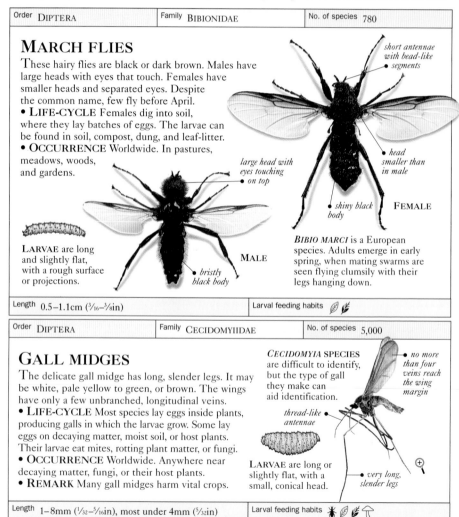

Order DIPTERA	Family BIBIONIDAE	No. of species 780

MARCH FLIES

These hairy flies are black or dark brown. Males have large heads with eyes that touch. Females have smaller heads and separated eyes. Despite the common name, few fly before April.
• **LIFE-CYCLE** Females dig into soil, where they lay batches of eggs. The larvae can be found in soil, compost, dung, and leaf-litter.
• **OCCURRENCE** Worldwide. In pastures, meadows, woods, and gardens.

short antennae with bead-like segments

large head with eyes touching on top

head smaller than in male

shiny black body

FEMALE

LARVAE are long and slightly flat, with a rough surface or projections.

MALE

bristly black body

BIBIO MARCI is a European species. Adults emerge in early spring, when mating swarms are seen flying clumsily with their legs hanging down.

Length 0.5–1.1cm (³⁄₁₆–³⁄₈in)	Larval feeding habits

Order DIPTERA	Family CECIDOMYIIDAE	No. of species 5,000

GALL MIDGES

The delicate gall midge has long, slender legs. It may be white, pale yellow to green, or brown. The wings have only a few unbranched, longitudinal veins.
• **LIFE-CYCLE** Most species lay eggs inside plants, producing galls in which the larvae grow. Some lay eggs on decaying matter, moist soil, or host plants. Their larvae eat mites, rotting plant matter, or fungi.
• **OCCURRENCE** Worldwide. Anywhere near decaying matter, fungi, or their host plants.
• **REMARK** Many gall midges harm vital crops.

***CECIDOMYIA* SPECIES** are difficult to identify, but the type of gall they make can aid identification.

no more than four veins reach the wing margin

thread-like antennae

LARVAE are long or slightly flat, with a small, conical head.

very long, slender legs

Length 1–8mm (¹⁄₃₂–⁵⁄₁₆in), most under 4mm (⁵⁄₃₂in)	Larval feeding habits

Order DIPTERA	Family CERATOPOGONIDAE	No. of species 4,000

BITING MIDGES

The common name of these pests comes from their habit of biting vertebrates or other insects. Slender or stocky, with short, strong legs, these flies are dull grey or brown, usually with dark wing mottling. The head is very rounded, and the male's feathery antennae are sensitive to the female's wing beats.
• **LIFE-CYCLE** Males and females mate while flying in a swarm. Eggs are laid in groups or strings in wet soil, rotting matter, bogs, and water. Adults do not fly very far from the boggy larval breeding areas and suck blood from a wide range of vertebrates. Some species suck body fluids from larger insects, while others catch and eat very small insects or eat a variety of other matter.
• **OCCURRENCE** Worldwide, but mainly in the northern hemisphere. Common by margins of ponds, rivers, and lakes, in bogs, and near seashores.
• **REMARK** The bites of these insects can produce severe irritation, making working outside almost impossible in some regions. In warmer areas, some biting midges transmit worm parasites to humans and some carry animal diseases such as African horse sickness and Bluetongue. These midges are also, however, important crop pollinators.

LARVAE are worm-like, with a distinct head. They may have hairs on the body.

small head in relation to thorax •

indistinct vein patterns, especially • at end of wings

CULICOIDES IMPUNCTATUS is notorious around Scottish lochs, streams, and boggy areas. The female's bite is extremely itchy and painful.

Length 1–5mm (¹⁄₃₂–³⁄₁₆in), most around 3mm (¹⁄₈in)	Larval feeding habits ✳ 🐛 🌱 💧

Order DIPTERA	Family CHIRONOMIDAE	No. of species 5,000

NON-BITING MIDGES

These humpbacked flies can be robust or delicate, with long, slender legs, and are pale brown or slightly green in colour. They look like mosquitoes but lack functional mouthparts.
• **LIFE-CYCLE** Most of the two- to three-year life-cycle is lived as larvae; adults live no longer than a couple of weeks. Mating occurs on the wing in a mating swarm, and eggs are laid in a mass of sticky jelly on water or plants. The larvae eat decayed organic matter, algae, and tiny plants or aquatic animals, but some are predacious or burrow into aquatic plants.
• **OCCURRENCE** Worldwide. Widespread in many habitats and often seen in swarms at dusk near ponds, lakes, and streams.
• **REMARK** The larvae form a vital part of aquatic food webs. Some have haemoglobin in their bodies, which helps them to live in stagnant, muddy water.

in this specimen, wings are twisted and so only edge is seen •

haltere •

• male has characteristically feathery antennae

slender body of male •

• long legs with fine hairs

long, slender legs

LARVAE are elongate and often have a pair of prolegs on the prothorax and the last abdominal segment.

△ ***CHIRONOMUS RIPARIUS*** is a European species whose larvae live in stagnant ponds and backwaters.

CHIRONOMUS **SPECIES** are fragile flies. The adults are short-lived – mostly just for a few days – but can be present in very large numbers.

Length 1–9mm (¹⁄₃₂–¹¹⁄₃₂in)	Larval feeding habits ✳ 🍃 🐛

Order DIPTERA	Family CULICIDAE	No. of species 3,100

MOSQUITOES

The body of these delicate, slender flies is covered with patterns of white, grey, brown, and black scales. Some tropical species are brightly coloured. The long, narrow wings have scales along the veins and edges, and the mouthparts are very long, slender, and piercing.

• LIFE-CYCLE Eggs are laid on the surface of water, either singly or in groups of 30 to 300. The larvae are saprophagous or eat other mosquito larvae and obtain air from the water surface through an abdominal siphon. Adult females suck the blood of vertebrates but also take plant fluids and nectar; the males feed on plant fluids, nectar, and honeydew. The life-cycle usually lasts fewer than three weeks.

• OCCURRENCE Worldwide, especially in warm areas. Adults are common near woodland, and larvae are found in virtually any body of fresh water, from a rain-filled treehole or water-butt to a lake.

• REMARK The females of many species carry organisms that cause serious diseases in animals and humans. These diseases include malaria, yellow fever, dengue fever, filariasis, and encephalitis. Malaria is caused by protozoa that belong to the genus *Plasmodium* and that are parasites of blood. Worldwide, one person dies of malaria every 12 seconds. Despite massive efforts to control the disease, it continues to be a serious, and growing, problem.

LARVAE are elongate, with tufts of hair on the large thoracic segments.

females have slightly feathery antennae

short palps

grey or whitish scales

dark brown coloration

haltere

pale bands on hind tarsi

abdomen covered with scales

△ *AEDES CANTANS* is a European species that breeds in temporary pools of water and bites humans.

long palps

characteristic feeding position: body in line with head pointed down towards skin

blood meal visible inside body

△ *ANOPHELES GAMBIAE* belongs to the main malaria-carrying genus of mosquitoes and is a widespread vector of the disease in Africa. It usually bites humans rather than animals.

short palps

long proboscis

banded abdomen

CULEX SPECIES rest with their head and abdomen pointing down towards the surface. Various species carry diseases, including filariasis.

Length 0.3–2cm (⅛–¾in), most under 1cm (⅜in)	Larval feeding habits

Order DIPTERA	Family MYCETOPHILIDAE	No. of species 3,300

FUNGUS GNATS

Small and mosquito-like, these flies are generally brown, black, or yellowish. However, some species may be brightly coloured. The thorax is characteristically humped, and the legs are long and slender.
• **LIFE-CYCLE** Eggs are usually laid in fungi, on or under bark, on cave walls, and in nests. The larvae mostly eat fungi and are found inside fungi, in dead wood and decaying vegetation, or under bark. Some larvae eat tiny insects and worms. Pupation occurs inside larval food or in a silk cocoon.
• **OCCURRENCE** Worldwide. In damp, dark places such as woods and caves, and near rotting vegetation and fungi.
• **REMARK** The cave-dwelling, luminescent larvae of *Arachnocampa* species, which live in Australia and New Zealand, catch flying insects on sticky threads that they hang from the cave rock.

long and slender antennae

very elongate first leg segment (coxa)

dark, slender abdomen

humped thorax

△ *MACROCERA STIGMA* is a widespread western European species with hairs on its wings.

spines on tibiae

LARVAE are white, smooth, and very slender, with a brown or black head.

long first segment of tarsi

PLATYURA MARGINATA is a fairly large, dark-coloured species, which is found in western Europe. Its larvae live underneath decaying wood.

Length 0.2–1.5cm (¹⁄₁₆–⅝in)	Larval feeding habits

Order DIPTERA	Family PSYCHODIDAE	No. of species 2,000

SAND FLIES AND MOTH FLIES

Small and moth-like, these grey to brown flies are covered with hairs or scales. The wings, usually broad with pointed tips, are held together over the body in sand flies and are tent-like or partly spread in moth flies. Most species are nocturnal.
• **LIFE-CYCLE** Sand flies feed on vertebrate blood, including that of humans, but the eggs are laid, and larvae develop, in wet soil. Moth flies lay eggs, and live, near wet, decaying matter.
• **OCCURRENCE** Worldwide, especially in warm areas. Sand flies favour warm, dry habitats; moth flies prefer woods near streams and bogs.
• **REMARK** Some sand flies spread disease, especially leishmaniasis in humans.

long hairs on wing margin *broad wings with pointed tips* *small body covered in hairs*

LARVAE are elongate, and some taper to the rear.

PERICOMA FULIGINOSA is a dark, widely distributed moth fly. Its larvae develop in shallow water and in tree rot holes.

Length 1–5mm (¹⁄₃₂–³⁄₁₆in)	Larval feeding habits

Order DIPTERA	Family SIMULIIDAE	No. of species 1,500

BLACK FLIES

Usually black or dark brown, these flies have a stout, hump-backed appearance. The antennae and legs are short.
• **LIFE-CYCLE** Eggs are laid in running water. Larvae stick to stones and plants by means of a holdfast organ at their rear and pupate in submerged cases. When the adults emerge, they rise to the surface of the water and fly off.
• **OCCURRENCE** Worldwide. Near flowing water.
• **REMARK** *Simulium* species carry the roundworm that causes river blindness in tropical regions.

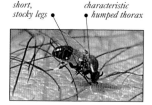

short, stocky legs • • *characteristic humped thorax*

LARVAE use their posterior sucker to anchor themselves.

SIMULIUM SPECIES are also known as black flies, although some are not entirely black. Many species carry disease.

Length 1–5mm (¹/₃₂–³/₁₆in), most under 4mm (⁵/₃₂in)	Larval feeding habits

Order DIPTERA	Family TIPULIDAE	No. of species 15,000

CRANE FLIES

Also called daddy-long-legs, these fragile flies are well known for shedding their very long legs easily if caught. They are mostly brown, black, or grey with yellow or pale brown markings. The end of the abdomen is blunt and expanded in males, while females have a pointed ovipositor.
• **LIFE-CYCLE** Eggs are typically laid in soil. The larvae live in soil, rotting wood, birds' nests, and bogs, where they eat roots, decaying organic material, fungal threads, and mosses. Some aquatic crane flies may be carnivorous. Many adults are short-lived, fly at twilight, and may feed on nectar.
• **OCCURRENCE** Worldwide. Often found by water or among damp vegetation, shaded woodland, or pasture.
• **REMARK** The larvae of many crane flies are pests of crops, garden plants, and lawns.

LARVAE have tough bodies and so are often called leatherjackets.

extremely long and fragile legs •

front of head elongate •

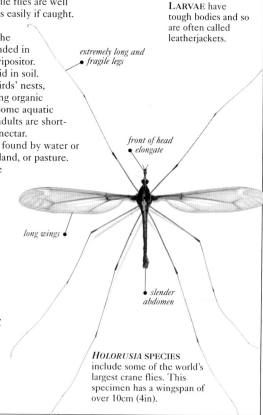

male antennae may appear feathery •

long wings •

• *smoky wing-tip patches*

• *black-and-yellow, wasp-like markings*

• *slender abdomen*

CTENOPHORA ORNATA is a distinctive European species whose larvae develop in well-decayed wood.

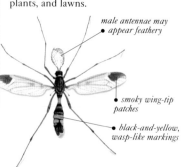

HOLORUSIA SPECIES include some of the world's largest crane flies. This specimen has a wingspan of over 10cm (4in).

Length 0.6–6cm (¼–2½in), most 1.2–2.4cm (½–1in)	Larval feeding habits

| Order DIPTERA | Family ACROCERIDAE | No. of species 500 |

SMALL-HEADED FLIES

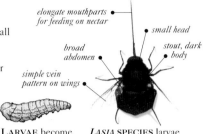

True to their name, the head of these flies is small and the eyes cover most of its area. The body is stout, and the thorax has a humped appearance.
• LIFE-CYCLE Eggs are laid on grass or twigs or dropped in flight. Hatched larvae seek out and parasitize young spiders. Once inside a spider's body, the larva does not develop further until the spider reaches its last moult. The larva itself then moults, eats the spider's internal organs, and leaves the spider's body in order to pupate.
• OCCURRENCE Worldwide. Various habitats, wherever spiders are found.

elongate mouthparts for feeding on nectar
small head
broad abdomen
stout, dark body
simple vein pattern on wings

LARVAE become fat and grub-like once inside a spider's body.

LASIA SPECIES larvae develop inside the bodies of tarantula spiders in South America.

| Length 0.3–2.2cm (⅛–⅞in) | Larval feeding habits |

| Order DIPTERA | Family AGROMYZIDAE | No. of species 2,500 |

LEAF-MINING FLIES

These flies are grey, black, or greenish yellow and may have patterned wings. The abdomen tapers, and females have a pointed ovipositor.
• LIFE-CYCLE Eggs are laid in plant tissue. Larvae chew "mines" (channels) through leaves or feed inside stems, seeds, or roots. Some form galls. Pupation occurs in the mine or in soil.
• OCCURRENCE Worldwide. Wherever their host plants occur.
• REMARK These flies are crop pests. A few species are used to control weeds.

stout, curved, black bristles on head and thorax
smoky tint on wings

LARVAE are white or pale yellow and slightly flat.

HEXOMYZA SPECIES are found in the UK, USA, Japan, and South Africa. Its larvae make galls on the twigs of trees such as poplars.

| Length 1–6mm (¹⁄₃₂–¼in) | Larval feeding habits |

| Order DIPTERA | Family ANTHOMYIIDAE | No. of species 1,500 |

ANTHOMYIID FLIES

Many anthomyiids look like yellowish, dull brown, grey, or black house flies (see p.148).
• LIFE-CYCLE Eggs are laid in or on plant tissue, and the larvae are found boring into stems, mining leaves, or inside galls on a huge range of host plants. Some develop in rotting seaweed or dung, and a few species live as parasites inside the nests of solitary bees and wasps.
• OCCURRENCE Worldwide, mainly in the northern hemisphere. In a wide range of wooded, damp habitats or near seashores.

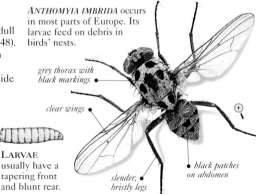

ANTHOMYIA IMBRIDA occurs in most parts of Europe. Its larvae feed on debris in birds' nests.

grey thorax with black markings
clear wings
black patches on abdomen
slender, bristly legs

LARVAE usually have a tapering front and blunt rear.

| Length 0.2–1.2cm (¹⁄₁₆–½in), most 0.7–0.9cm (⁹⁄₃₂–¹¹⁄₃₂in) | Larval feeding habits |

Order DIPTERA	Family ASILIDAE	No. of species 5,000

ROBBER FLIES

The head of these slender or bee-like flies is characteristically slightly hollow between the eyes, with a long tuft of hairs on the face.
• **LIFE-CYCLE** Robber flies stab insect prey through a weak point, such as the neck, and inject saliva. The contents of the paralysed insect's body are then sucked up. Eggs are laid either in soil or on or inside plants. Most larvae live in soil, leaf-litter, or rotting wood, where they eat the eggs, larvae, and pupae of other insects.
• **OCCURRENCE** Worldwide. In a variety of habitats, especially dry or semi-arid grasslands.

sharp, forward-pointing proboscis

long tuft of facial hairs

stout, bristly legs

broad wingspan

hairy body in bee-mimicking species

LARVAE are long, cylindrical, and often pointed at both ends.

smoky tinge at wing-tips

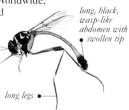

△ *BLEPHAROTES SPLENDIDISSIMUS* is found in Australia. This large species has plate-like tufts of hair at the sides of its flat abdomen.

PAGIDOLAPHRIA FLAMMIPENNIS has a long, sideways-flattened proboscis. Some of the facial hairs are as long as the proboscis.

Length 0.3–7cm (⅛–2¾in), most 0.8–1.5cm (⁵⁄₁₆–⅝in)	Larval feeding habits 🐜 🌿

Order DIPTERA	Family BOMBYLIIDAE	No. of species 5,000

BEE FLIES

Although some can be small, most bee flies tend to be stout and hairy, hence their common name. Many species are brown, red, and yellow in coloration, and some have bright markings.
• **LIFE-CYCLE** The larvae of most known bee flies parasitize the larvae of various other insects, although a few eat grasshopper eggs. Females produce many small eggs, which may, for example, be laid near the nest of a host bee. The active first-stage larvae of the bee fly will then locate the host bee larvae in their nest, eat them, and proceed to pupate inside the bee's cell. Adult bee flies feed on nectar.
• **OCCURRENCE** Worldwide, especially in open and semi-arid regions. Around flowers or resting on the ground.

proboscis, for sucking nectar, is two to three times longer than head

stout, hairy body

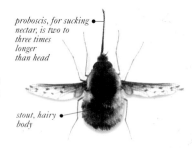

△ *BOMBYLIUS DISCOLOR* is a European species. With its broad abdomen and furry body, it looks very much like a bumblebee.

long, black, wasp-like abdomen with swollen tip

long legs

LARVAE are curved and narrow towards both ends.

SYSTROPUS SPECIES are slim-bodied, extremely wasp-like insects from tropical and subtropical regions.

mottled wings

brightly marked body

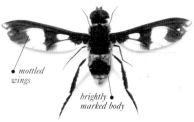

LIGYRA VENUS is a distinctively patterned species from Tanzania. Its larvae develop inside the nests of certain wasps.

Length 0.2–3cm (¹⁄₁₆–1¼in), most under 2cm (¾in)	Larval feeding habits 🐜 🐝

Order DIPTERA	Family CALLIPHORIDAE	No. of species 1,200

BLOW FLIES

These flies are typically stout and may be metallic green or blue, shiny black, or dull. In some species, the sexes are of different colours. This family includes the familiar bluebottles and greenbottles.

• LIFE-CYCLE Blow flies lay eggs on carrion, dung, and flesh. The larvae of certain species are predators of ants, termites, and other insect larvae and eggs, and a few suck the blood of nestlings. Some blow flies lay larvae rather than eggs.

• OCCURRENCE Worldwide. On flowers, vegetation, and carcasses. Also attracted to cooked and raw food.

• REMARK Many blow flies lay their eggs on livestock and humans and carry disease. The sheep maggot fly, *Lucilia sericata*, for example, lays eggs on the wool of sheep, and its larvae burrow into the flesh. A few blow-fly species burrow into human flesh and have been used in surgery as a way of removing dead tissue.

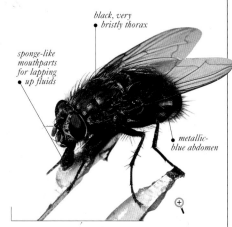

black, very bristly thorax

sponge-like mouthparts for lapping up fluids

metallic-blue abdomen

CALLIPHORA VOMITORIA is a bluebottle species that is extremely common in the countryside. Its females may lay many hundreds of eggs during their lifetime.

LARVAE are white or pale. They taper at the front and are blunt at the rear. Bands of tiny spines encircle the body.

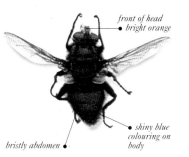

front of head bright orange

bristly abdomen

shiny blue colouring on body

◁ *CYNOMYIA MORTUORUM* is a common bluebottle whose larvae develop in rotten meat, corpses, and human excrement.

▽ *CALLIPHORA VICINA* is more common in towns and cities, where the maggots develop inside the corpses of dead animals such as pigeons, rats, and mice.

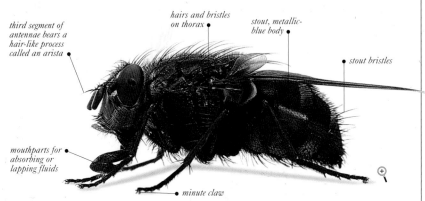

third segment of antennae bears a hair-like process called an arista

hairs and bristles on thorax

stout, metallic-blue body

stout bristles

mouthparts for absorbing or lapping fluids

minute claw

Length 0.4–1.5cm (⅛–⅝in)	Larval feeding habits

Order DIPTERA	Family CELYPHIDAE	No. of species 100

BEETLE FLIES

These flies owe their common name to the beetle-like hind part of the thorax (the scutellum). Hugely enlarged and often metallic-coloured, this covers the abdomen and the folded wings.
• LIFE-CYCLE Eggs are laid, and the larvae develop, in decaying vegetation.
• OCCURRENCE Tropical regions, except Central and South America. In damp habitats near water and in grasslands.

enlarged rear part of thorax

wings folded under thorax

eye

antennae

LARVAE are small, pale, and maggot-like. They taper at both ends.

CHAEMAECELYPHUS SPECIES look like small beetles. They push their wings out from under the rear of the thorax to fly.

Length 3–8mm (⅛–⁵⁄₁₆in)	Larval feeding habits

Order DIPTERA	Family CHLOROPIDAE	No. of species 2,000

GRASS FLIES

Also known as stem flies, these common insects are blackish grey, green, or black with yellow markings. There is a distinct triangular mark on the top of the head.
• LIFE-CYCLE Eggs are laid mostly on or in plant tissue, or whatever the larvae eat. Most larvae bore into grasses. Others form galls or eat decaying plant matter, root aphids, or the eggs of spiders and other insects.
• OCCURRENCE Worldwide. Widespread in a range of habitats, from grasslands to rainforests.
• REMARK Several species attack cereal crops and some cause blindness in humans and animals.

simple vein pattern on wings

green body

dark stripes on thorax

mouthparts

curved tibiae on stout hindlegs

LARVAE are usually slender, blunt at the rear, and narrow at the front.

MEROMYZA PRATORUM is found in coastal sand dunes, where its larvae tunnel into the stems of marram grass.

Length 1–6mm (¹⁄₃₂–¼in), most under 4mm (⁵⁄₃₂in)	Larval feeding habits

Order DIPTERA	Family CONOPIDAE	No. of species 1,000

THICK-HEADED FLIES

These flies have broad heads and an abdomen that narrows where it joins the thorax. Many resemble bees or wasps in shape and coloration.
• LIFE-CYCLE Eggs are laid on the bodies of hosts such as other flies, crickets, cockroaches, wasps, or bees. Larvae burrow inside and feed on body fluids.
• OCCURRENCE Worldwide. In varied habitats, usually feeding at flowers.
• REMARK Some larvae may cause bee hosts to burrow into the soil before they die, giving the flies a safe place to pupate.

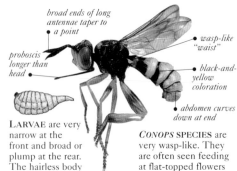

broad ends of long antennae taper to a point

proboscis longer than head

wasp-like "waist"

black-and-yellow coloration

abdomen curves down at end

LARVAE are very narrow at the front and broad or plump at the rear. The hairless body has fine wrinkles.

CONOPS SPECIES are very wasp-like. They are often seen feeding at flat-topped flowers such as hogweed.

Length 0.3–2.6cm (⅛–1in)	Larval feeding habits

Order DIPTERA	Family DIOPSIDAE	No. of species 180

STALK-EYED FLIES

The most recognizable feature of these small flies is the head. This is extended on both sides into stalks that bear the eyes and the antennae. The eye stalks are larger in the males and may be absent in the females.

• **LIFE-CYCLE** Eggs are stuck on to young foliage or rotting vegetation. Males engage in combat over territories and females. Those with more widely spaced eyes tend to have more success in finding mates.

• **OCCURRENCE** Tropical regions, especially Africa. Absent from South America, and only one very short-stalked species occurs in North America. Often found near or on vegetation or close to running water.

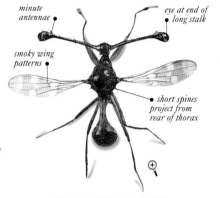

minute antennae

eye at end of long stalk

smoky wing patterns

short spines project from rear of thorax

LARVAE taper at both ends and have smooth, hairless bodies.

CYRTODIOPSIS DALMANNI is found in parts of Southeast Asia. Its larvae bore into the stems of various grass species.

Length 0.3–1.8cm (⅛–¾in)	Larval feeding habits

Order DIPTERA	Family DOLICHOPODIDAE	No. of species 5,500

LONG-LEGGED FLIES

These small, bristly flies are metallic-bronze, -green, or -blue. The rounded head has a short, fleshy proboscis, and the legs are distinctively long and slender.

• **LIFE-CYCLE** Usually, eggs are laid, and the larvae are found, under bark and in wet soil, water, mud, leaf-litter, and seaweed. The larvae mostly eat other insects, while the adults eat soft-bodied insects or feed on nectar.

• **OCCURRENCE** Worldwide. In damp meadows, woods, and streams and at lakesides and seashores.

• **REMARK** Some larvae are predators of pests such as the larvae of bark beetles or mosquitoes.

smoky patches on wings

brilliant metallic-green body

male genitalia

long, pale legs

bristly tibiae

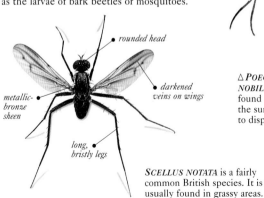

rounded head

darkened veins on wings

metallic-bronze sheen

long, bristly legs

△ *POECILOBOTHRUS NOBILITATUS* males are often found on patches of bare mud, in the sun. They wave their wings to display them to females.

SCELLUS NOTATA is a fairly common British species. It is usually found in grassy areas.

LARVAE are pale and taper slightly to the front. The rear of the abdomen often has four lobes.

Length 0.1–1cm (1⁄32–⅜in), most under 4mm (5⁄32in)	Larval feeding habits

Order DIPTERA	Family DROSOPHILIDAE	No. of species 2,900

POMACE FLIES

curved bristles on back

striped abdomen

bright red eyes

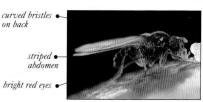

Also called lesser fruit flies or vinegar flies, these small yellow, brown, or black species usually have light or bright red eyes. The thorax and the abdomen may be striped or spotted.
• **LIFE-CYCLE** Eggs are laid in or near a food supply: usually bacteria and other micro-organisms, but also fungi, carrion, or dung. A few species burrow inside leaves or eat spittle-bug nymphs (see p.95) or aquatic fly larvae.
• **OCCURRENCE** Worldwide. On rotting vegetation and fruit, or near fungi or fermenting liquids.
• **REMARK** Half of all pomace flies are in the genus *Drosophila*.

△ **DROSOPHILA** SPECIES are often attracted to rotting fruit. The one shown here is feeding on a piece of apple.

translucent wings

◁ **DROSOPHILA MELANOGASTER** is the best known of the pomace flies. It is often used for genetics studies as it breeds quickly and has large chromosomes in its salivary glands.

LARVAE have hooked spines all around each segment.

brownish coloration

Length 1–6mm (1/32–1/4in)	Larval feeding habits ✳ 🍃 🦟

Order DIPTERA	Family EMPIDIDAE	No. of species 3,500

DANCE FLIES

The common name of this family derives from the dancing motions of mating swarms. Most species have a stout thorax and a tapering, elongate abdomen. The rounded head has large eyes and usually a sharp, downward-pointing proboscis.
• **LIFE-CYCLE** Courtship and mating may involve males offering prey as food to females. Eggs are usually laid on soil, dung, rotting wood or plant matter, leaf-litter, and water. The larvae develop in these places and feed on small insect prey. Adults eat small flies and drink nectar.
• **OCCURRENCE** Worldwide, mainly in the northern hemisphere. On vegetation in damp habitats; in "dancing" swarms above water.

pale halteres

distinctive antennae

stout, piercing proboscis

slender abdomen

long, bristly legs

LARVAE are slender and aquatic species have prolegs.

WIEDEMANNIA STAGNALIS adults are predacious and sit on wet moss waiting for suitable prey. The larvae probably develop inside clumps of wet moss.

EMPIS SPECIES have a large, downward-pointing proboscis. This enables the fly to "stab" its prey and extract the fluids on which it feeds.

Length 0.15–1.1cm (1/32–1/2in)	Larval feeding habits ✳

Order DIPTERA	Family GASTEROPHILIDAE	No. of species 50

HORSE BOT FLIES

dark patches on wings

These stout flies resemble honeybees (see pp.180–81). They have non-functional mouthparts and do not feed. All horse bot species are internal parasites of large mammals such as horses, rhinoceroses, and elephants. Some species may parasitize humans.

• **LIFE-CYCLE** The short-lived adults lay eggs on grass or near the host's mouth. The larvae burrow into the host or are swallowed. They live in the host's gut and, when mature, are passed out with the excrement to pupate in the soil.

short antennae

• **OCCURRENCE** Worldwide, especially in Asia and Africa. Near host animals.

hairy, bee-like body

LARVAE are thick-bodied, with distinct bands of backward-pointing spines.

GASTEROPHILUS INTESTINALIS is a parasite of horses. It flies around the horse and darts in to lay eggs on the skin, which are swallowed when the host grooms itself.

Length 1–2.5cm (⅜–1in)	Larval feeding habits

Order DIPTERA	Family GLOSSINIDAE	No. of species 22

TSETSE FLIES

GLOSSINA MORSITANS is one of the species that transmits the trypanosome parasite – the cause of sleeping sickness.

These brown or grey flies feed on animal or human blood and cause sleeping sickness in humans and nagana in animals. At rest, they cross their wings over their abdomen.

• **LIFE-CYCLE** Females produce eggs singly. The larva hatches inside the female's body and feeds on glandular secretions. Once deposited on the ground, it pupates immediately and the adult emerges after four weeks.

needle-like mouthparts

stout body

LARVAE are as large as the adults when mature.

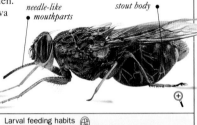

• **OCCURRENCE** Africa. In wooded savannah and bush.

Length 0.6–1.4cm (¼–⅝in)	Larval feeding habits

Order DIPTERA	Family HIPPOBOSCIDAE	No. of species 200

LOUSE FLIES

strong, clawed legs

These stout, flat flies have a short proboscis and strong, clawed legs for gripping hair or feathers. They are parasites, feeding on the blood of mammals and birds.

head appears partly sunk into thorax

• **LIFE-CYCLE** Larvae develop inside the female. Laid when mature, the larvae pupate on their hosts.

• **OCCURRENCE** Worldwide. On host animals, including cattle, sheep, horses, deer, and birds.

LARVAE are white or yellow. When fully grown, they are fat and round.

CRATAERINA PALLIDA is a parasite of swifts. Three-quarters of all louse fly species parasitize birds.

vestigial wings

Length 0.15–1.2cm (¹⁄₃₂–½in)	Larval feeding habits

Order DIPTERA	Family MUSCIDAE	No. of species 3,500

HOUSE FLIES

Most house flies are drably coloured, with dark bristles
and long, slender legs. The mouthparts are sponge-like,
for lapping fluids, or piercing, for sucking blood.
• **LIFE-CYCLE** Masses of eggs are laid in excrement,
rotting matter, fungi, birds' nests, water, or plants. The
larvae (maggots) grow fast and can pupate in just over a week.
• **OCCURRENCE** Worldwide. On flowers, excrement, and
rotting matter; blood-sucking species are found near their hosts.
• **REMARK** House flies may carry infections, including
typhoid and cholera.

• *fine
bristles*

• *orangish
patches at side
of abdomen*

• *large eyes*

MUSCA DOMESTICA breeds
in fermenting organic matter
and spreads a variety of
diseases, including cholera,
typhoid, dysentery, and
gastroenteritis.

LARVAE mostly
taper at the head and
are blunt at the rear.

Length 0.2–1.2cm (¹⁄₁₆–½in)	Larval feeding habits

Order DIPTERA	Family MYDIDAE	No. of species 350

MYDAS FLIES

Many of these flies mimic spider-hunting
wasps and some mimic hoverflies (see
p.153 and pp.188–89). They are
typically smooth-bodied and
black, although the tips of the
wings and antennae may be
orange. Adult mouthparts are weak.
The males are often territorial.
• **LIFE-CYCLE** Females lay their
eggs in soil or rotting wood. The larvae
have large, curved mandibles and eat
beetle larvae and insect pupae.
• **OCCURRENCE** Worldwide,
especially in warmer regions of
Africa and South America. In
sandy, dry areas or woodland.
• **REMARK** This family
contains the world's
largest flies. The larvae
of *Mydas heros* live inside
the nests of leaf-cutter
ants, where they eat
scarab beetle larvae.

*last segment of long
antennae swollen or*
• *clubbed*

LARVAE may be
cylindrical and taper
at the head end.

• *dark wings
have a wrinkled
appearance*

• *dark, wasp-like
appearance*

MYDAS HEROS looks
like a dark wasp. It is a
large, strong-legged fly
from South America.

strong legs •

Length 2–6cm (¾–2⅜in)	Larval feeding habits

Order DIPTERA	Family NYCTERIBIIDAE	No. of species 250

BAT FLIES

These small, spidery, wingless flies prey on the blood of bats. They have a narrow head, and the eyes are either tiny or totally lacking.
• **LIFE-CYCLE** The eggs hatch, and larvae develop, inside the mother's abdomen, fed by special internal glands. Pupae are laid in the bat roost. Adults emerge when they sense a bat nearby.
• **OCCURRENCE** Worldwide. In or near bat roosts.

for feeding, head hinges forwards from groove at back of thorax

unusually, legs join back of distorted thorax

LARVAE are soft, fat, and white and darken as they mature.

NYCTERIBIA KOLENATII is found on several bat species and shows the distorted thorax that is typical of bat flies.

Length 2–4mm (1⁄16–5⁄32in)	Larval feeding habits

Order DIPTERA	Family OESTRIDAE	No. of species 80

BOT AND WARBLE FLIES

Many of these heavy, large-headed, hairy flies look like stout-legged bees. The short-lived adults do not feed.
• **LIFE-CYCLE** The larvae are all internal parasites of mammals such as sheep and cattle. Bot flies lay larvae directly into the nostrils of their hosts. Warble flies lay eggs on the hair of the host, and the larvae burrow under the skin. Fully grown larvae emerge and pupate in the soil.
• **OCCURRENCE** Worldwide, especially in the northern hemisphere and Africa. Near their hosts.

small antennae, in grooves on head

large head

hairy, heavy body

LARVAE are fat, taper at the head, and have bands of spines.

HYPODERMA BOVIS, or the Ox Warble Fly, is a widespread species. It may look like a bee, but cattle recognize it and twist, jump, or run to get out of its way.

Length 0.8–2.5cm (5⁄16–1in)	Larval feeding habits

Order DIPTERA	Family PANTOPTHALMIDAE	No. of species 25

TIMBER FLIES

These flies look like large horseflies (see p.154) but with small, non-biting mouthparts. The abdomen is typically square and the wings darkly patterned.
• **LIFE-CYCLE** Small groups of eggs are laid in cracks in tree bark, and the larvae burrow into the wood. It is uncertain whether they digest the wood or eat sap, fungi, and decaying matter.
• **OCCURRENCE** South America. In rainforests.

antennae narrow at apex

eyes do not meet in females

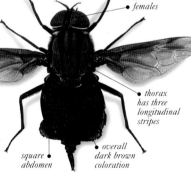

LARVAE are stout and may take up to two years to mature.

PANTOPHTHALMUS BELLARDII is widespread in tropical South America. The adults of this dark brown species probably feed on sap and other fluids.

thorax has three longitudinal stripes

square abdomen

overall dark brown coloration

Length 0.3–5cm (1⁄8–2in)	Larval feeding habits

Order DIPTERA	Family PHORIDAE	No. of species 3,000

SCUTTLE FLIES

These small brown, black, or yellowish flies are also known as hump-backed flies because of their distinctive appearance. They have a small, strongly down-turned head and hind femora that are often flat and highly enlarged. The body bristles look feathery under magnification.

• LIFE-CYCLE Eggs are laid, and larvae develop, in a wide range of microhabitats. Some larvae feed on fungi, carrion, or decaying matter; others are scavengers or parasitic on other insects, snails, millipedes, or worms.

• OCCURRENCE Worldwide. In a wide variety of habitats.

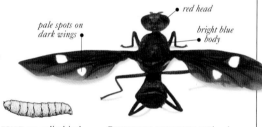

small head, turned sharply downwards

strongly developed thorax

strong hindlegs

LARVAE are fattest in the middle, often with spiny projections.

ANEVRINA THORACICA is native to the northern hemisphere. Its larvae favour soil, the corpses of small animals, and moles' nests.

Length 0.6–6mm (¹⁄₄₄–¹⁄₄in)	Larval feeding habits

Order DIPTERA	Family PLATYSTOMATIDAE	No. of species 1,200

SIGNAL FLIES

Many of these flies have bright coloration and patterned wings. Their antennae lie within grooves on the head, and the males' eyes may be on stalks.

• LIFE-CYCLE Eggs are laid on all kinds of decaying matter. Adults mate around trees and foliage, and males use their wings to display.

• OCCURRENCE Worldwide, especially in warm, humid parts of Europe, Asia, and Africa. In a wide range of habitats.

pale spots on dark wings

red head

bright blue body

LARVAE are cylindrical and blunt at the rear. Short spines under the abdomen help with locomotion.

CLITODOCA FENESTRALIS is a large African species with a wingspan of 4.5cm (1¾in). Its bright coloration is probably important in courtship.

Length 0.4–2cm (¹⁄₃₂–¾in)	Larval feeding habits

Order DIPTERA	Family PSILIDAE	No. of species 250

RUST FLIES

These slender flies are reddish brown to black, with a slightly triangular or rounded head. The common name refers to the rust-red, flaking appearance of plant roots that have been affected by the larvae.

• LIFE-CYCLE Eggs are laid on host plants or in soil near the roots. The larvae of most rust flies bore through plant stems or roots and under bark; some are gall-formers.

• OCCURRENCE Mainly in the northern hemisphere. In woodland and damp areas.

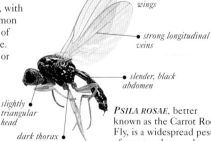

translucent wings

strong longitudinal veins

slender, black abdomen

slightly triangular head

LARVAE are pale, smooth, slender, and cylindrical.

dark thorax

PSILA ROSAE, better known as the Carrot Root Fly, is a widespread pest of carrot, celery, and parsnip plants.

Length 3–9mm (¹⁄₈–¹¹⁄₃₂in)	Larval feeding habits

Order DIPTERA	Family SARCOPHAGIDAE	No. of species 2,500

FLESH FLIES

These flies are mostly a dull, silvery grey or black. The thorax is longitudinally striped, and the abdomen looks chequered or marbled.
• **LIFE-CYCLE** The common name refers to the fact that some flesh flies lay larvae in body cavities and on wounds in vertebrates, including man. Most females give birth to live first-stage larvae. They either lay these larvae or drop them in flight and retain the egg shell within their body. Some larvae feed on carrion, while others are parasitic on other insects, snails, worms, or other invertebrates.
• **OCCURRENCE** Worldwide, especially in the northern hemisphere. In varied habitats.

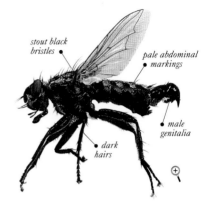

stout black bristles

pale abdominal markings

male genitalia

dark hairs

LARVAE have pointed heads, blunt rears, and bands of fine spines.

SARCOPHAGA MELANURA is a British species, found near coastal areas. Its larvae are found in rotting matter and may parasitize snails or insects.

Length 0.2–2cm (1⁄16–3⁄4in), most 0.6–1cm (1⁄4–3⁄8in)	Larval feeding habits

Order DIPTERA	Family SCATHOPHAGIDAE	No. of species 350

DUNG FLIES

These flies are generally dull grey, brown, or yellow-brown but may be black or yellow and black. Their slender legs may have strong, dark bristles. The most common dung flies are very hairy. The common name is misleading as it refers only to flies of the genus *Scathophaga*.
• **LIFE-CYCLE** Typically, eggs are laid, and the larvae develop, in plants, where the young eat foliage and may be leaf-miners, or in dung, where they eat other larvae. Some larvae are found in damp soil and water, where they prey on small invertebrates. Adults catch and eat smaller insects.
• **OCCURRENCE** Northern hemisphere. In various habitats, including all kinds of plants and fresh dung.

SCATHOPHAGA STERCORARIA, also called the Yellow Dung Fly, is widespread across the northern hemisphere. Common on sheep and cow dung, it also breeds on the dung of poultry, horses, and humans.

hairy body

slender legs

yellowish brown coloration of male (females are greenish)

strong, dark bristles on legs

LARVAE are pale and cylindrical. Some taper to a point at the head end.

Length 0.3–1.2cm (1⁄8–1⁄2in)	Larval feeding habits

Order DIPTERA	Family SEPSIDAE	No. of species 250

BLACK SCAVENGER FLIES

Also known as ensign flies, due to the fact that males signal to females by waving their wings, most of these flies are shiny black or brownish, and males often have dark wing-tips. Some small species are ant-like. The front legs of males are modified with bristles and spines so that they can grip the wings of females before and during mating.

• LIFE-CYCLE Eggs are laid – and the slim larvae are found – in mammals' dung, compost, and rotting meat on carcasses. Larvae may occur in huge numbers.

• OCCURRENCE Worldwide. In a wide range of habitats, on flowers and around dung or decaying organic matter.

• REMARK Some form huge swarms on plants at the end of the summer. It is thought that they mark places they have been with special odours and find them again after hibernation, when mating occurs.

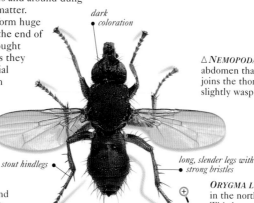

wing-tips tinged weakly with brown

antennae

narrow wings

almost spherical head

constricted abdomen

△ *NEMOPODA NITIDULA* has an abdomen that narrows where it joins the thorax, giving a slightly wasp-like appearance.

dark coloration

stout hindlegs

long, slender legs with strong bristles

LARVAE are narrow and taper towards the front. The last abdominal segment is swollen.

strong bristles on abdomen

ORYGMA LUCTUOSA is found in the northern hemisphere. This large species lays its eggs on rotting seaweed along shorelines.

Length 2–6mm (¹⁄₁₆–¼in)	Larval feeding habits

Order DIPTERA	Family STRATIOMYIDAE	No. of species 2,000

SOLDIER FLIES

Most of these flies are quite stout and slightly flat, with bright or metallic markings. Some are fairly large and wasp-like. The broad abdomen is typically banded yellow, black, or green, while the head is very rounded. In males, the head is largely covered by the eyes. Certain species are aquatic.

• LIFE-CYCLE Eggs are laid on the surface of water, on plants, or in dung, leaf-litter, soil, or rotten wood. Most larvae eat rotting matter or the larvae of flies and bark beetles.

• OCCURRENCE Worldwide. In a wide variety of habitats, often at flowers. Some aquatic species can even survive in hot springs or very salty water.

rounded head

very large eyes

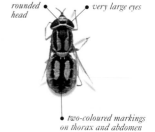

two-coloured markings on thorax and abdomen

LARVAE are toughened and flat, with many short bristles.

HEDRIODISCUS PULCHER is a native of South America. It has green coloration and distinctive markings.

Length 0.3–2cm (⅛–¾in)	Larval feeding habits

Order DIPTERA	Family SYRPHIDAE	No. of species 6,000

HOVER FLIES

Also called flower flies, these flies are highly distinctive as they hover and dart between flowers. Many species are quite slender and wasp-like, with yellow stripes, spots, or bands; others are stout, hairy, and bee-like. Some are black, blue, or metallic. A false vein runs down the centre of the wings, which also have false margins.

LARVAE are variable in form. Those that live in water or liquid manure typically have a breathing tube at the rear.

• **LIFE-CYCLE** Eggs are laid where the larvae feed (see below). Adults feed on pollen and nectar but the larvae have diverse feeding habits. The larvae of many hover flies eat huge numbers of aphids during the course of their development. Other predatory species attack scale insects, sawfly larvae, and soft-bodied insects. Some larvae feed in or on decaying wood, dung, mud, or stagnant water, and a few feed on plants or fungi. Certain larvae live inside the nests of bees and social wasps, where they eat the dead larvae and pupae of their hosts.

• **OCCURRENCE** Worldwide. In a variety of habitats, usually on flat-topped flowers.

• **REMARK** Although a few herbivorous species are pests in bulbs, most are important plant pollinators.

SERICOMYIA SILENTIS is found on acid heathland. The larvae of this species live in boggy pools such as those that form after cutting peat.

large eyes of males usually meet on top of head ●

● black-and-yellow legs

small, dark wing patches ●

● broad wedges of yellow at each side of abdomen

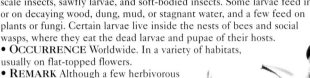

wasp-like shape and colouring ●

● short antennae

false vein ●

false margin ●

△ *SYRPHUS RIBESII* is a common, wasp-like species found in Europe. It often gathers in large numbers, and its larvae eat greenflies.

round head ●

eyes in females well separated ●

front of wings orange-yellow ●

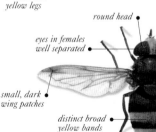

● rear of wings pale, with false margin

distinct broad yellow bands ●

VOLUCELLA ZONARIA is a stout, distinctively banded hover fly. It is a native of Europe and is migratory. The larvae of this species scavenge inside wasps' nests.

Length 0.3–3.4cm (⅛–1¼in), most 1–2cm (⅜–¾in)	Larval feeding habits

Order DIPTERA	Family TABANIDAE	No. of species 4,000

HORSE FLIES

Also called deer flies, clegs, and gad flies, these stout, hairless insects have colourful, patterned eyes in flat, round heads. Most are black, grey, or brown with broad abdomens that often have bright bands or markings. The females' blade-like mouthparts are adapted for cutting skin. Male horse flies do not have biting mouthparts and drink at pools and flowers.

• LIFE-CYCLE Eggs are laid in places such as soil and rotting wood. Larvae typically live in wet soil or mud near ponds and streams, where they eat worms, crustaceans, and insect larvae. Some live in rotting tree-holes or decaying wood. Adults feed on pollen and nectar. Females also take blood from mammals and birds.

• OCCURRENCE Worldwide. In a very wide range of habitats, near mammals.

• REMARK In warm regions, horse flies can spread diseases affecting animals and humans.

large eyes

round head, concave behind

TABANUS ATRATUS is a black horse fly found throughout the USA. This and other species bite cattle and can damage beef production. They may also transmit viruses to livestock.

TABANUS SPECIES are dark and robust. This large specimen is probably mimicking a big bee such as a carpenter bee (see p.179)

LARVAE are mostly tough and slightly shiny, with fine, longitudinal striations.

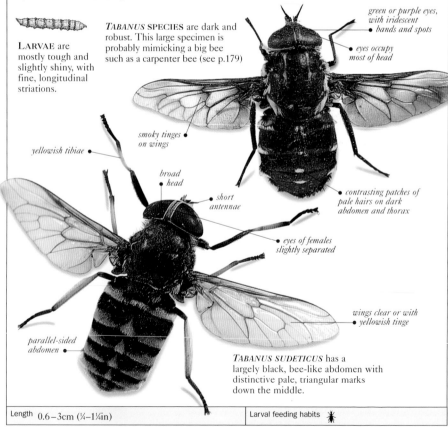

green or purple eyes, with iridescent bands and spots

eyes occupy most of head

yellowish tibiae

smoky tinges on wings

broad head

short antennae

eyes of females slightly separated

contrasting patches of pale hairs on dark abdomen and thorax

parallel-sided abdomen

wings clear or with yellowish tinge

TABANUS SUDETICUS has a largely black, bee-like abdomen with distinctive pale, triangular marks down the middle.

Length 0.6–3cm (¼–1¼in)	Larval feeding habits

Order DIPTERA	Family TACHINIDAE	No. of species 8,000

PARASITIC FLIES

These stout flies are very variable in appearance. Many species look like bristly house flies (see p.148), while some larger species can look almost bee-like. The abdomen is especially bristly, particularly towards the rear end.

• **LIFE-CYCLE** The larvae are mostly parasitic on insects. Eggs are laid either on the host, with the hatched larvae burrowing inside, or inside the host. Females of some species lay their eggs directly into the mouths of feeding insects or on plants that the hosts will eat.

• **OCCURRENCE** Worldwide. In a wide variety of habitats, wherever their hosts are found.

• **REMARK** Many species are used as biological control agents.

head appears much narrower than abdomen

stout, orangish body

thick black spines or bristles on abdomen

△ *PARADEJEANIA RUTILOIDES* is found in northwestern parts of the USA. It attacks and parasitizes several different types of caterpillar.

LARVAE are white or yellowish, perhaps with spines or hairs.

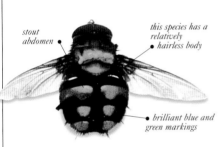

stout abdomen

this species has a relatively hairless body

brilliant blue and green markings

FORMOSIA MONETA is a stout fly with highly distinctive coloration. It attacks and parasitizes the larvae of scarab beetles (see pp.128–29).

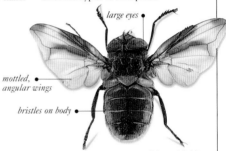

large eyes

mottled, angular wings

bristles on body

PHASIA HEMIPTERA parasitizes shield bugs (see p.92). It is found in meadows and woodland in parts of Europe, including the United Kingdom.

Length 0.5–1.5cm (³⁄₁₆–⁵⁄₈in)	Larval feeding habits

Order DIPTERA	Family TEPHRITIDAE	No. of species 4,500

FRUIT FLIES

Most fruit flies have distinctive wing patterns, which can be used to identify individual species. These can take the form of bands, patches, and zig-zag markings. Females have a pointed ovipositor that may be longer than the rest of the body.

• **LIFE-CYCLE** Eggs are laid on and in plants. The larvae of some species feed inside soft fruits or flowerheads, while others are leaf-miners or gall-formers.

• **OCCURRENCE** Worldwide. In a wide variety of habitats.

• **REMARK** Many species are crop and fruit pests, for example *Ceratitis capitata*, the Mediterranean Fruit Fly, which damages citrus and other soft fruits.

orange thorax and head

mottled pattern on wings

very dark abdomen

LARVAE vary in shape. The body may be smooth or slightly spiny.

ICTERICA WESTERMANNI is a European species of fruit fly that feeds on the flowerheads of ragworts.

Length 0.2–2cm (¹⁄₁₆–³⁄₄in), most under 1.5cm (⁵⁄₈in)	Larval feeding habits

CADDISFLIES

M EMBERS of the order Trichoptera, containing 43 families and 8,000 species, are found almost anywhere there is fresh water. The slender, dull adults look very moth-like, but unlike moths their body and wings are covered with hairs, not scales. The long, thin antennae are multi-segmented, and the weakly developed mouthparts may be used to take liquid, although the adults of many species do not feed. Caddisflies have compound eyes, sometimes accompanied by ocelli. In flight, the hindwings and forewings are coupled by curved hairs.

Females typically lay masses or strings of jelly-encased eggs below the surface of water, attached to plants. Metamorphosis is complete. The aquatic larvae usually pupate inside cases that they make from materials such as sand grains and twigs.

Order TRICHOPTERA	Family HYDROPSYCHIDAE	No. of species 1,000

NET-SPINNING CADDISFLIES

These caddisflies are drably coloured with either hairy or clear wings. There are wart-like projections on the pronotum.
• **LIFE-CYCLE** Eggs are laid in water. The stout larvae live near a cup-shaped net that they spin between stones or other objects. The net catches small organisms, algae, and debris, which the larvae eat. Some larvae are predacious.
• **OCCURRENCE** Worldwide. Common along streams and rivers.
• **REMARK** Different species make nets with varying mesh sizes to suit their preferred food.

drab, mottled colouring • *hairy head* • *long antennae* •

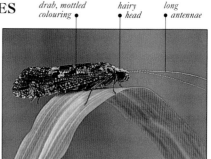

LARVAE have well-branched gills on the thorax and abdomen.

HYDROPSYCHE CONTUBERNALIS is native to various parts of western Europe. It is found in running rather than still water. This species has the drab coloration typical of caddisflies.

Length 0.6–1.8cm (¼–¾in)	Larval feeding habits

Order TRICHOPTERA	Family HYDROPTILIDAE	No. of species 1,000

MICRO-CADDISFLIES

Also known as purse case-makers, these caddisflies have black, white, or grey speckled coloration. They are densely covered with hair.
• **LIFE-CYCLE** Eggs are laid in jelly-like masses, either on water or on water plants. The first four larval stages are active and suck the juices of water plants. The last larval stage makes an open-ended purse or barrel-shaped case out of silk from its salivary glands.
• **OCCURRENCE** Worldwide. Near streams, rivers, ponds, and lakes.

long, multi-segmented antennae • *tuft of hairs on head* • *drab coloration* •

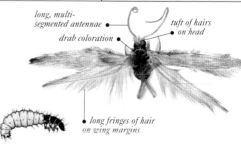

long fringes of hair on wing margins •

LARVAE are small and free-living in the early stages.

HYDROPTILA **SPECIES** are found all over the world – this is the largest, most widespread genus in the family, containing 150 species.

Length 2–6mm (¹⁄₁₆–¼in)	Larval feeding habits

| Order TRICHOPTERA | Family LIMNEPHILIDAE | No. of species 1,500 |

NORTHERN CADDISFLIES

These species are dark brown or slightly red or yellow. The wings have dark markings, a straight front margin, and appear to be "cut-off" at the rear. The front legs each have a tibial spur.
• LIFE-CYCLE Eggs are laid in water. The larvae make cases that often look like tiny, irregular log cabins. Most larvae eat organic detritus, algae, and other small organisms.
• OCCURRENCE Mainly in the northern hemisphere. Around ponds, lakes, streams, ditches, temporary pools, and marshes.

drab brown coloration

papery wings with few hairs

LARVAE have a round head and may be quite large.

LIMNEPHILUS LUNATUS is a widespread species. Its larvae are found in a wide variety of freshwater habitats. Coloration is variable, but is generally in drab shades of black and brown.

| Length 0.7–3.0cm (⁹⁄₃₂–1¼in), most under 2.4cm (1in) | Larval feeding habits |

| Order TRICHOPTERA | Family PHILOPOTAMIDAE | No. of species 500 |

FINGER-NET CADDISFLIES

These are small and darkly coloured species with oval wings. They have fairly flat heads with ocelli.
• LIFE-CYCLE Eggs are laid in water. The larvae live inside finger-shaped, fine-meshed nets that they attach to rocks. The mouthparts are used to brush up filtered organic particles.
• OCCURRENCE Worldwide. Especially common near fast-flowing streams.

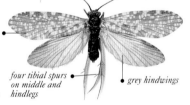

yellow-and-brown patterned forewings

four tibial spurs on middle and hindlegs

grey hindwings

LARVAE have a red-tinged head and pronotum.

PHILOPOTAMUS MONTANUS, as the name suggests, is usually found around fast-flowing streams in hilly or mountainous parts of Europe.

| Length 4–8mm (⁵⁄₃₂–⁵⁄₁₆in) | Larval feeding habits |

| Order TRICHOPTERA | Family PHRYGANEIDAE | No. of species 450 |

LARGE CADDISFLIES

Members of this family have light brown or grey markings and may look mottled. Ocelli are present. There are at least two tibial spurs on the front legs and four on the middle and hindlegs.
• LIFE-CYCLE Eggs are laid in water. The larvae make light cases of spirally arranged plant fragments and fibres, adding material as they grow.
• OCCURRENCE Mainly northern hemisphere. Near ponds, lakes, bogs, and slow-flowing streams and rivers.

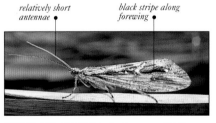

relatively short antennae

black stripe along forewing

LARVAE are slim and flat. The head may have dark bands.

PHRYGANEA GRANDIS is the largest caddisfly found in the UK. The male is smaller than the female (shown here) and lacks the distinctive dark stripe along the forewing.

| Length 1.2–2.6cm (½–1in) | Larval feeding habits |

MOTHS AND BUTTERFLIES

T HE 127 FAMILIES AND 165,000 species of moths and butterflies make up the order Lepidoptera. There is no scientific difference between moths and butterflies. Moths, however, usually fly at night and butterflies during the day, while butterflies have club-ended antennae, which moths usually lack. Both groups have tiny, overlapping scales on the body and wings and multi-segmented antennae. The mouthparts usually form a proboscis for taking nectar and other liquids. The selection below places moths first (Arctiidae–Zygaenidae) and butterflies second (Lycaenidae–Pieridae).

Courtship involves displays and odours. Either sex releases a scent that is carried downwind and picked up by the mate's antennae. Eggs are scattered or laid on the larval food plants. Metamorphosis is complete. The cylindrical larvae (caterpillars), the majority of which are herbivorous, have chewing mouthparts, three pairs of thoracic legs, and a variable number of abdominal prolegs, with tiny hooks to grip food plants. There are four to nine larval stages. The pupa (chrysalis) may be: underground in a silk-lined cell; surrounded by a silk cocoon produced by the mature larva; or naked and attached to the food plant.

Order LEPIDOPTERA	Family ARCTIIDAE	No. of species 2,500

TIGER AND ERMINE MOTHS

Most tiger moths are heavy-bodied and hairy, often brightly coloured in various combinations of black, red, yellow, and orange, which warns off predators. Ermine moths tend to be pale or white, with small black markings. This family also includes the slender, dull footmen moths.
• **LIFE-CYCLE** Eggs are laid on and around host plants. Many caterpillars in this family eat a wide range of plant matter and the majority of species feed at night. They are very hairy, and many species are poisonous because they eat the leaves of plants such as potato and laburnum, which contain toxic substances. Footmen species feed by day, and their smooth caterpillars eat lichens.
• **OCCURRENCE** Worldwide. In well-vegetated areas where host plants occur.
• **REMARK** Some are significant forest and orchard pests.

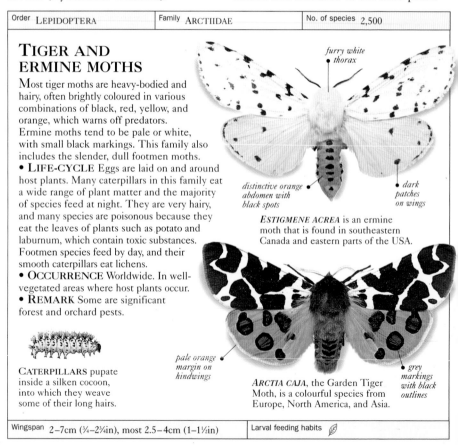

furry white thorax

distinctive orange abdomen with black spots

dark patches on wings

ESTIGMENE ACREA is an ermine moth that is found in southeastern Canada and eastern parts of the USA.

CATERPILLARS pupate inside a silken cocoon, into which they weave some of their long hairs.

pale orange margin on hindwings

ARCTIA CAJA, the Garden Tiger Moth, is a colourful species from Europe, North America, and Asia.

grey markings with black outlines

Wingspan 2–7cm (¾–2¾in), most 2.5–4cm (1–1½in)	Larval feeding habits 🍃

Order LEPIDOPTERA	Family BOMBYCIDAE	No. of species 100

SILK MOTHS

These stout, hairy moths are pale cream, grey, or brown. They have no working mouthparts and do not feed.
• LIFE-CYCLE Eggs are laid on host plants. Pupation occurs in a silk cocoon. *Bombyx mori* caterpillars eat mulberry leaves; other species eat a variety of plants.
• OCCURRENCE Southeast Asia. In well-vegetated areas wherever their host plants proliferate.

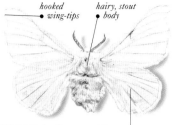

hooked • wing-tips
hairy, stout • body
• conspicuous wing veins

CATERPILLARS are smooth, with prolegs on some abdominal segments.

BOMBYX MORI, the Commercial Silk Moth, originated in Asia. It is now found in silk farms worldwide.

Wingspan 2–6cm (¾–2½in)	Larval feeding habits

Order LEPIDOPTERA	Family BRAHMAEIDAE	No. of species 20

BRAHMAEID MOTHS

These large moths have wavy wing patterns. The forewings may have eye-spots.
• LIFE-CYCLE Eggs are laid on host plants, and pupation takes place on the ground.
• OCCURRENCE Eastern Europe, Asia, and Africa. In forests and woods, where host trees occur.

BRAHMAEA WALLICHII, the Owl Moth, is an Eastern species and one of the largest in this family.

CATERPILLARS are often colourful, with abdominal processes.

• large eye-spots

Wingspan 5–16.5cm (2–6½in)	Larval feeding habits

Order LEPIDOPTERA	Family CASTNIIDAE	No. of species 180

CASTNIID MOTHS

These day-flying, very butterfly-like moths have distinctive, broad wings. Their forewings are usually coloured for camouflage but the hindwings can be either brightly coloured or metallic, with white or orange spots or bands.
• LIFE-CYCLE Eggs are laid on host plants. The caterpillars are stem-borers or feed on roots.
• OCCURRENCE Central and South America, Southeast Asia, and Australia. In well-vegetated areas, wherever their host plants occur.

antennae are • clubbed and hooked at the tips

CATERPILLARS are pale, hairless, and grub-like.

CASTNIA LICUS, the Giant Sugar-cane Borer, is also a significant pest of bananas.

Wingspan 3–11cm (1¼–4¼in)	Larval feeding habits

Order LEPIDOPTERA	Family COSSIDAE	No. of species 700

CARPENTER MOTHS

These heavy-bodied species usually have spotted, drab wings, irregularly patterned brown, white, and cream. Some species are called leopard moths, as they have black-spotted white wings.
• **LIFE-CYCLE** Females lay eggs on bark or in the tunnels from which the moths emerge as adults. The caterpillars mainly tunnel and eat woody tissue. Carpenter moths take from one to four years to reach adulthood. When fully grown, caterpillars pupate in their tunnels or in soil, inside a cocoon of silk and chewed wood fibres.
• **OCCURRENCE** Worldwide. In woodland areas.
• **REMARK** Some species are pests of oak, maple, pine, and other trees. "Witchety" grubs, eaten by aboriginal people, are carpenter moth larvae.

large black spots on
white thorax

white wings with
black spots

ZEUZERA PYRINA, the Leopard Moth, is found across the northern hemisphere.

pale head
and collar

CATERPILLARS are fat-bodied and are commonly known as carpenter worms.

distinctly banded
abdomen

COSSUS COSSUS is the European Goat Moth. If disturbed, its larvae can produce a foul smell.

Wingspan 2–22.5cm (¾–8¾in)	Larval feeding habits

Order LEPIDOPTERA	Family DREPANIDAE	No. of species 1,000

HOOK-TIP MOTHS

These moths are so-called because many have hooked tips on their forewings. Most species are slender-bodied, with drably coloured wings.
• **LIFE-CYCLE** Female hook-tip moths lay flat eggs on host plants – the caterpillars feed on the foliage of shrubs and trees. In the caterpillars of many species, the clasping prolegs at the end of the abdomen are extremely reduced, and the tail-end may appear either tapered or pointed.
• **OCCURRENCE** Tropical regions except South America. In woodland and well-vegetated areas.

male has feathery
antennae

distinctive hooked tip
on forewings

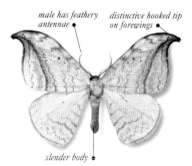

slender body

CATERPILLARS of some species rest with the head and tail raised or the front of the body curved around.

DREPANA ARCUATA, also known as the Arched Hook-tip, occurs in North America. Its caterpillars feed on the foliage of birch and alder trees.

Wingspan 2–5cm (¾–2in)	Larval feeding habits

Order LEPIDOPTERA	Family GEOMETRIDAE	No. of species 20,000

GEOMETER MOTHS

The wings of these slender-bodied moths are rather large and rounded with complex patterns of fine markings. They are generally nocturnal and usually have brown or green camouflage colouring. Some tropical species are brightly coloured day-fliers.
• LIFE-CYCLE Eggs are laid singly or in groups on the bark, twigs, or stems of host plants, and hatch in spring. When disturbed, the caterpillars of many species remain still and look twig-like. When fully grown, caterpillars spin a fragile cocoon between leaves or in litter.
• OCCURRENCE Worldwide. Almost anywhere vegetation is found.
• REMARK Many species are agricultural and forestry pests, and can cause severe damage or defoliation.

• broad wings

• scalloped wing margins

GEOMETRA PAPILIONARIA, or Large Emerald, caterpillars feed on the foliage of beech, alder, and hazel trees.

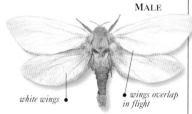

• chequered wing fringes

RHEUMAPTRA HASTATA, known as the Argent and Sable Moth, is found across the northern hemisphere.

CATERPILLARS are commonly called inchworms or loopers because of their distinctive looping motion.

Wingspan 1.4–7.4cm (½–3in)	Larval feeding habits

Order LEPIDOPTERA	Family HEPIALIDAE	No. of species 300

GHOST MOTHS

Also called swift moths, due to their fast flight, most species have similarly shaped fore- and hindwings. Many have drab general coloration, while some have patterns of bright silver spots. The wings are not joined in flight, as in other moths, but simply overlap.
• LIFE-CYCLE Females drop eggs singly on the ground, by host plants, and can produce several hundred eggs. The caterpillars mainly bore into plant stems, trunks, and roots, but a few eat leaves or moss.
• OCCURRENCE Worldwide, especially in Southeast Asia and Australia. In open woodland and grassland.
• REMARK Many species eat a range of plant matter and can become pests of grasses, vegetables, shrubs, and trees.

MALE

white wings •

• wings overlap in flight

HEPIALUS HUMULI caterpillars are yellow-white with small, dark spots. They can become harmful pests of potato, lettuce, and strawberry crops.

pinkish brown • forewing pattern

CATERPILLARS of this family are white, with a brown head.

FEMALE

fore- and hindwings similarly shaped •

• furry abdomen

Wingspan 3–24cm (1¼–9½in)	Larval feeding habits

Order LEPIDOPTERA	Family HESPERIIDAE	No. of species 3,000

SKIPPERS

Although closely related to butterflies, these day-flyers are typically moth-like and heavy-bodied. The antennae end in a club, which is long, curved, and pointed. While the forewings are usually short and triangular, the hindwings may be tailed. The common name refers to their rapid, darting flight.

• LIFE-CYCLE Females lay single eggs on their host plant. The caterpillars of most species feed at night on grasses and sedges, herbaceous plants, and the foliage of some trees. During the day, they hide inside a shelter of rolled or folded leaves.

• OCCURRENCE Worldwide, except in New Zealand. In a variety of open habitats, such as cultivated fields and grassland.

dark-bordered red patch • orange bands on thorax • elongate club at end of antennae

△ *AMENIS BARONI* is a brightly coloured species from Peru, with highly distinctive orange-red forewing markings.

yellow-orange hindwing margins

CATERPILLARS are green, brown, or white. They usually have a large head with a distinct neck and taper towards the rear.

pale blue patch •

yellow • patches

triangular forewings •

yellow bands • red tail •

△ *EUSCHEMON RAFFLESIA*, the vivid Regent Skipper, lives in the rainforests of Australia and is seen feeding during the day at nectar-rich flowers.

• drab brown wings with white markings

white patches on wings •

• iridescent green hairs on body

long tails on hindwings •

△ *URBANUS PROTEUS* is a common species in North and South America. Its long, tailed hindwings make it highly recognizable.

▷ *CALPODES ETHLIUS*, the Brazilian Skipper, is widespread in South America and the West Indies. This large and robust skipper often flies great distances.

distinct lobe • on hindwing

Wingspan 2–8cm (¾–3¼in), most under 4.5cm (1¾in)	Larval feeding habits

Order LEPIDOPTERA	Family INCURVARIIDAE	No. of species 300

INCURVARIID MOTHS

Most of these small moths have camouflage colouring. A few are a metallic gold or bronze. Some species with long antennae are called fairy moths.
• **LIFE-CYCLE** Single eggs are laid inside plant tissue. The caterpillars are seed-borers or leaf-miners in their early stages and make a case out of plant material, inside which they live and eventually pupate.
• **OCCURRENCE** Worldwide, except in New Zealand. In woods, wherever their host plants grow.

shiny wing scales

abdomen tapers sharply

NEMOPHORA CUPRIACELLA has metallic wing scales, an orange tuft of hair on its head, and white antennal tips.

▽ *NEMOPHORA SCABIOSELLA*, from Asia and Europe, has dark wing fringes.

very long antennae

CATERPILLARS are small, with tiny abdominal prolegs.

Wingspan 0.8–2.5cm (⅕6–1in)	Larval feeding habits

Order LEPIDOPTERA	Family LASIOCAMPIDAE	No. of species 2,000

LAPPET AND EGGAR MOTHS

Most of these very hairy, heavy-bodied moths are yellowish brown, brown, or grey in colour. Females are bigger than males and generally have large abdomens.
• **LIFE-CYCLE** Eggs are laid on host plants. The caterpillars live communally in silk tents or webs, spun across the foliage of various trees, grasses, and plants. Pupation occurs inside tough, papery, egg-like cocoons.
• **OCCURRENCE** Worldwide, except in New Zealand. Anywhere their host trees and plants occur.

pale-bordered band across middle of forewings

CATERPILLARS are stout with tufts of hairs on both their back and sides.

MALACOSOMA AMERICANUM, the Eastern Tent Moth, is a troublesome pest of apple and wild cherry trees.

Wingspan 2.5–9.5cm (1–3¾in)	Larval feeding habits

Order LEPIDOPTERA	Family LIMACODIDAE	No. of species 1,000

LIMACODID MOTHS

Most limacodids have broad, rounded wings, hairy bodies, and dull colouring. The name means "slug-like" and refers to the shape and locomotion of the caterpillars.
• **LIFE-CYCLE** Flat eggs are laid on the leaves of host plants. The caterpillars are often poisonous or are armed with stinging hairs.
• **OCCURRENCE** Worldwide, especially in tropical regions. On a wide range of shrubs and trees.

dull-coloured, hairy body

two white spots on forewings

CATERPILLARS are often brightly coloured. They have no prolegs.

SIBINE STIMULEA, the Saddle-back Moth of North America, is named after the saddle-shaped mark on the back of its caterpillars.

Wingspan 2–4.5cm (¾–1¾in)	Larval feeding habits

Order LEPIDOPTERA	Family LYMANTRIIDAE	No. of species 2,600

TUSSOCK MOTHS

These moths look similar to noctuid moths (see p.165), but they are more hairy. Most are dull-coloured, but the tropical species can be colourful. Males are slightly smaller than the females, which are sometimes wingless. Adults lack a proboscis – and so do not feed.

• LIFE-CYCLE Females lay their eggs in batches on the bark of host trees and shrubs, often incorporating some of the irritant hairs from the end of the abdomen as a protective device against predators. The caterpillars, which may be brightly coloured, feed in groups on foliage.

• OCCURRENCE Worldwide. In many habitats, including hedgerows and coniferous or deciduous woods.

• REMARK The Gypsy and Brown-tail moths are serious pests of a range of trees and shrubs across the northern hemisphere. Outbreaks of these species can cause great damage, defoliating large areas and killing trees.

CATERPILLARS are very hairy – commonly with brush-like tufts of hairs on their back and sides.

• males have feathery antennae

MALE • tuft of hairs

LYMANTRIA DISPAR, the Gypsy Moth, is a native of Europe and Asia but was introduced to North America to produce inexpensive silk. However, the moth escaped and became a serious pest.

distinctive V-shaped • mark on forewings

• black dots on wing margins

broad, hairy body • FEMALE

pure white wings • with no markings

white, crescent-shaped spot •

only the male has • wings

• hairy hindwing margins

• large tuft of irritant hairs used to cover and protect eggs

ORGYIA ANTIQUA, otherwise known as the Vapourer Moth, is found in the northern hemisphere. The females are wingless.

EUPROCTIS CHRYSORRHOEA, the Brown-tail Moth, has hairy, brown caterpillars that live in a communal silk nest.

Wingspan 2–6cm (¾–2½in)	Larval feeding habits

Order LEPIDOPTERA	Family NOCTUIDAE	No. of species 22,000

NOCTUID MOTHS

These medium-sized nocturnal moths have fairly narrow forewings and broad hindwings. Noctuid moths have basically dull colouring, although the hindwings of some species are brightly coloured and patterned.
• **LIFE-CYCLE** Females lay eggs singly or in groups, at the base of host plants or in the soil. Caterpillars feed after dark and most attack their host plants, chewing or boring their way inside.
• **OCCURRENCE** Worldwide. In most habitats.
• **REMARK** These nocturnal moths have thoracic hearing organs for detecting bats. Many species are serious pests, between them damaging almost all the world's important crops. One species (see right), evolved from a fruit-piercing moth, sucks blood.

pointed wing-tips

CALYPTRA EUSTRIGATA, or the Vampire Moth, sucks blood and has barbed mouthparts that it uses to pierce the skin of mammals. It is found in India and Southeast Asia.

CATERPILLARS of most species feed on their host plants at night.

forewing has camouflage colouring and dark fringe

black, hairy body

tiny black dots along margin of forewings

irregular brown band

HELIOTHIS ARMIGERA, or the Old World Bollworm, is a serious pest of cotton, maize, and tomatoes. It is found across the eastern hemisphere.

XANTHOPASTIS TIMAIS, the Spanish Moth, is found in tropical North and South America. Its caterpillars feed on narcissus and fig species.

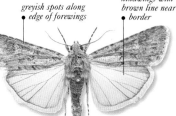

greyish spots along edge of forewings

translucent hindwings with brown line near border

distinctive black wing markings

SPODOPTERA EXIGUA is also known as the Small Mottled Willow Moth. It has a worldwide distribution and is a serious pest of cotton, maize, and rice.

dark, wavy line near margin of hindwings

AGROTIS IPSILON, the Dark Sword-grass Moth, is found throughout the world. Its caterpillars attack cotton, potatoes, tomatoes, and other crops.

Wingspan 1.5–16cm (⅝–6in), most under 8cm (3¼in)	Larval feeding habits ✳ 🍄 🍂 🐛 🍃

Order LEPIDOPTERA	Family NOTODONTIDAE	No. of species 3,000

PROMINENTS

Most prominents are drably coloured, with camouflage patterning. The common name refers to tufts of scales that, in some species, stick up prominently from the rear margins of the forewings when folded. A few species, such as the Buff-tip (*Phalera bucephala*), mimic broken twigs.
• LIFE-CYCLE Eggs are laid on the leaves of host plants. The caterpillars eat foliage and feed in groups to protect themselves from attacks by birds. Some produce chemicals and adopt threatening postures. Certain prominent caterpillars are described as "processionary", due to their night-time habit of moving in a long, head-to-tail line when seeking food. During the day, these species often shelter *en masse*, sometimes in a loose, silk nest.
• OCCURRENCE Worldwide. In a variety of habitats on their host plants – most commonly shrubs, trees, and leguminous plants.

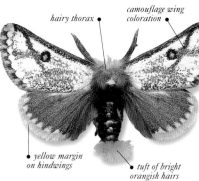

hairy thorax •

camouflage wing coloration

• yellow margin on hindwings

• tuft of bright orangish hairs

EPICOMA MELANOSTICA is an Australian species whose caterpillars feed on the leaves of *Leptospermum* species, common in southern and eastern parts of Australia.

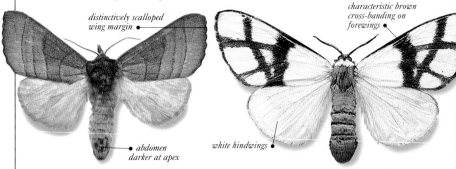

distinctively scalloped wing margin •

• abdomen darker at apex

DATANA MINISTRA is a rather drably coloured species, found in North America. Its caterpillars have black and yellow stripes and feed on a variety of deciduous tree foliage.

characteristic brown cross-banding on forewings •

white hindwings •

ANAPHE PANDA, the Banded Bagnest, lives in Africa and is named after the silk shelters spun by its caterpillars during the day. Some genus members defoliate their host trees.

dark spots at base and along edge of forewings

greyish brown zig-zag markings

• long, white, hair-like scales at wing bases

CERURA VINULA, the Puss Moth, is common in Europe and parts of Asia and has highly distinctive patterning.

• stout, furry abdomen

CATERPILLARS are brightly coloured, hairy, or striped, with fleshy bumps on their back.

Wingspan 3–8cm (1¼–3¼in)	Larval feeding habits

Order LEPIDOPTERA	Family PYRALIDAE	No. of species 24,000

SNOUT MOTHS

These moths are typically drably coloured. In some species, the front of the head appears to have a short "snout" formed by the long, sensory palps, held out straight. The forewings are broad or narrow, while the hindwings are broad and rounded. The legs are usually long.
• **LIFE-CYCLE** Eggs are laid near or on host plants, other host material, or prey. The larval feeding habits are diverse, but caterpillars typically burrow inside, and feed on, the leaves, stems, and roots of host plants. Some family members are scavengers, while a few prey on small insects and some even breed in sloth droppings or animal horn.
• **OCCURRENCE** Worldwide. In a wide range of habitats on their host plants.
• **REMARK** A large number of snout moths are pests of crops and dried fruit.

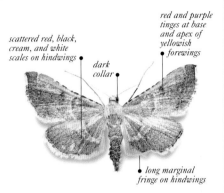

scattered red, black, cream, and white scales on hindwings

dark collar

red and purple tinges at base and apex of yellowish forewings

long marginal fringe on hindwings

△ *ENDOTRICHA FLAMMEALIS* is a nocturnal species. It is native to the United Kingdom and various parts of western Europe.

CATERPILLARS are either slender and cylindrical or quite stout. There are prolegs on some of the abdominal segments.

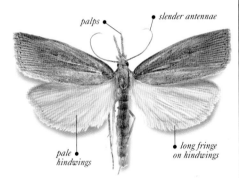

palps

slender antennae

pale hindwings

long fringe on hindwings

▷ *CHILO PHRAGMITELLA* is found in reedbeds, where its caterpillars feed on the stems of reeds belonging to the genus *Phragmites* – hence this moth's scientific name.

VITESSA SURADEVA is a bright, unusually patterned species from Borneo – all members of this genus are native to Southeast Asia.

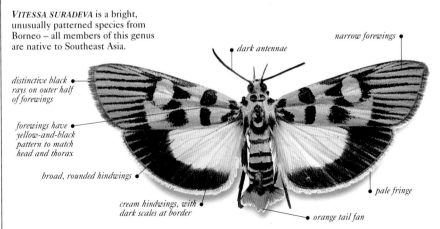

dark antennae

narrow forewings

distinctive black rays on outer half of forewings

forewings have yellow-and-black pattern to match head and thorax

broad, rounded hindwings

cream hindwings, with dark scales at border

orange tail fan

pale fringe

Wingspan 1–4.5cm (⅜–1¾in)	Larval feeding habits

Order LEPIDOPTERA	Family SATURNIIDAE	No. of species 1,200

SATURNIID MOTHS

Also known as emperor, moon, royal, and atlas moths, these large, heavy-bodied species have broad, often conspicuously marked wings. The mouthparts are entirely non-functional, and the adults do not feed. Antennae are feathery in the males and usually thread-like in the females. Species of the genus *Attacus*, found in Southeast Asia, are the largest moths in the world in terms of wing area, although the biggest moth wingspan is that of the Giant Agrippa Moth (*Thysania agrippina*), which belongs to the family Noctuidae (see p.165).

• LIFE-CYCLE Eggs are laid on a wide range of trees and shrubs, and the caterpillars feed on the foliage. Fully grown caterpillars make dense cocoons attached to the twigs of their host plants.

• OCCURRENCE Worldwide. Especially in wooded tropical and subtropical areas. *Attacus* species are protected in some countries, but many die by fluttering endlessly around street lighting.

• REMARK A few species of saturniid moth can be pests of various trees. Silk from the cocoons of many species has been used consistently in the past, although not to such a great extent as that of the Silk Moth (*Bombyx mori*, see p.159).

CATERPILLARS can grow very large and have fleshy outgrowths (scoli) with spines and long hairs.

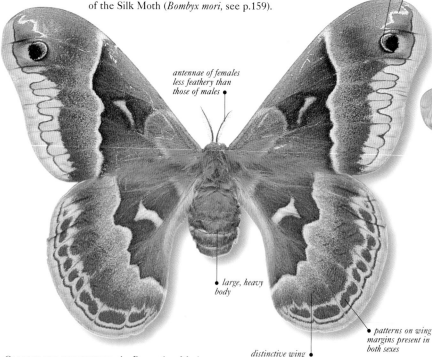

white zig-zag marking

white crescent around dark spot

antennae of females less feathery than those of males

large, heavy body

distinctive wing patterning, lacking in males

patterns on wing margins present in both sexes

CALLOSAMIA PROMETHEA, the Promethea Moth, is found in North America. The caterpillars feed on a range of foliage, including that of various fruit trees. Males are mostly black-brown with a pale border; females are a bright red-brown or dark brown and have pale wing markings.

Wingspan 5–30cm (2–12in)	Larval feeding habits

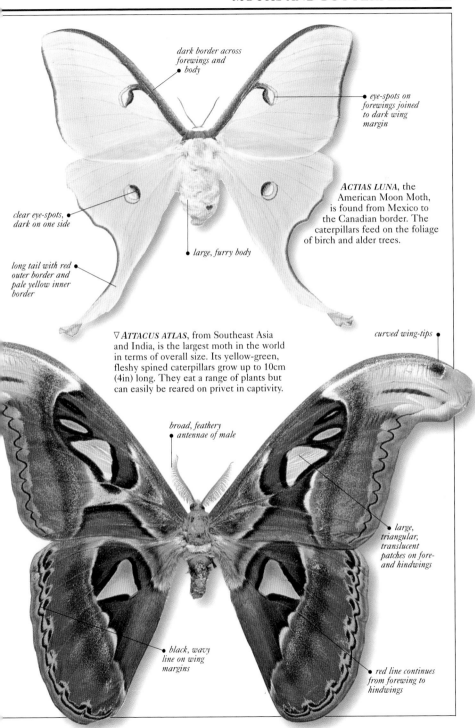

dark border across forewings and body

eye-spots on forewings joined to dark wing margin

ACTIAS LUNA, the American Moon Moth, is found from Mexico to the Canadian border. The caterpillars feed on the foliage of birch and alder trees.

clear eye-spots, dark on one side

long tail with red outer border and pale yellow inner border

large, furry body

▽ **ATTACUS ATLAS**, from Southeast Asia and India, is the largest moth in the world in terms of overall size. Its yellow-green, fleshy spined caterpillars grow up to 10cm (4in) long. They eat a range of plants but can easily be reared on privet in captivity.

curved wing-tips

broad, feathery antennae of male

large, triangular, translucent patches on fore- and hindwings

black, wavy line on wing margins

red line continues from forewing to hindwings

Order LEPIDOPTERA	Family SESIIDAE	No. of species 1,000

CLEAR-WINGED MOTHS

Some of these moths closely resemble wasps or bees. Their bodies are mainly black, bluish, or dark brown with yellow and orange markings, and their abdomens are often banded. They produce a buzzing sound during flight, which further emphasizes the mimicry. Some species even pretend to sting. Large areas of the wings are transparent, dark scales being present only along the veins. The end of the abdomen may have a fan-shaped tuft of long scales, and the antennae often have expanded or clubbed ends.

• **LIFE-CYCLE** Female clear-winged moths typically lay eggs on the trunks, stems, or roots of trees and shrubs, and the caterpillars burrow inside. When the moths first emerge from their pupae, their wings are fully covered in scales; the bare patches appear during the first flight.

• **OCCURRENCE** Worldwide. Around flowers or near their host plants.

• **REMARK** Many species, such as the Peach Tree Borer (*Synanthedon exitiosa*), are pests of fruit and other trees and shrubs.

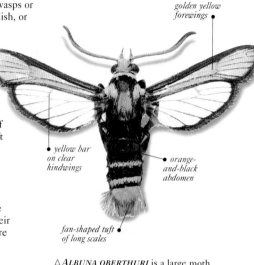

golden yellow forewings

yellow bar on clear hindwings

orange-and-black abdomen

fan-shaped tuft of long scales

△ *ALBUNA OBERTHURI* is a large moth from Australia's Northern Territory. It has distinctive hindwings – clear but with a noticeable yellow bar at the front – and an obvious abdominal tuft.

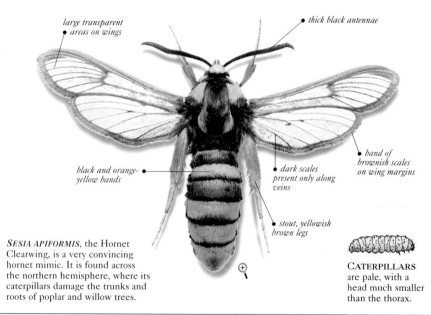

large transparent areas on wings

thick black antennae

black and orange-yellow bands

dark scales present only along veins

band of brownish scales on wing margins

stout, yellowish brown legs

SESIA APIFORMIS, the Hornet Clearwing, is a very convincing hornet mimic. It is found across the northern hemisphere, where its caterpillars damage the trunks and roots of poplar and willow trees.

CATERPILLARS are pale, with a head much smaller than the thorax.

Wingspan 1.5–4cm (⅝–1½in)	Larval feeding habits

Order LEPIDOPTERA	Family SPHINGIDAE	No. of species 1,100

HAWK MOTHS

These moths have an elongate, robust body and long, relatively narrow forewings. The long or very long proboscis is curled under the head when not in use. At rest, the wings are held back at a distinctive angle. Adults are strong fliers and most suck flower nectar. Some hover at flowers like hummingbirds.
• **LIFE-CYCLE** Single eggs are laid on plants, and the caterpillars eat their foliage.
• **OCCURRENCE** Worldwide, especially in tropical and subtropical regions. In a wide variety of habitats, wherever suitable host plants are found.

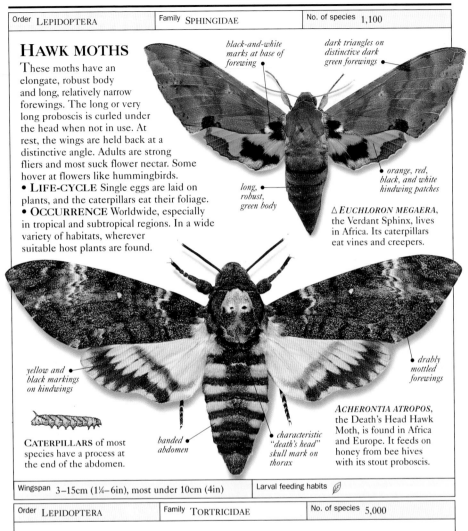

black-and-white marks at base of forewing

dark triangles on distinctive dark green forewings

orange, red, black, and white hindwing patches

long, robust, green body

△ *EUCHLORON MEGAERA*, the Verdant Sphinx, lives in Africa. Its caterpillars eat vines and creepers.

yellow and black markings on hindwings

drably mottled forewings

CATERPILLARS of most species have a process at the end of the abdomen.

banded abdomen

characteristic "death's head" skull mark on thorax

ACHERONTIA ATROPOS, the Death's Head Hawk Moth, is found in Africa and Europe. It feeds on honey from bee hives with its stout proboscis.

Wingspan 3–15cm (1¼–6in), most under 10cm (4in)	Larval feeding habits 🌿

Order LEPIDOPTERA	Family TORTRICIDAE	No. of species 5,000

TORTRICID MOTHS

These small moths are mostly brown, green, or grey, to blend in with bark, lichen, and leaves. Some species are brightly coloured. The forewings are broadly rectangular.
• **LIFE-CYCLE** Eggs are laid on host plants, including certain grasses, on which the caterpillars feed and pupate.
• **OCCURRENCE** Worldwide. In a wide range of habitats.
• **REMARK** Mexican "jumping beans" are the seeds of certain plants containing the wriggling larvae of a particular tortricid species.

pale hindwings with long fringe

forewings have camouflage colouring

CATERPILLARS may be slender or stout, and are relatively hairless.

CLEPSIS RURINANA is found in Europe and Asia. Its caterpillars may be found inside the rolled-up leaves of deciduous trees.

Wingspan 0.8–3cm (⁵⁄₁₆–1¼in)	Larval feeding habits 🌿

Order LEPIDOPTERA	Family URANIIDAE	No. of species 100

URANIID MOTHS

This family includes both large, long-tailed, colourful, day-flying moths with iridescent wing scales, and nocturnal species with dull coloration and without tails.

• LIFE-CYCLE Eggs are laid on host plants. The caterpillars of many species eat poisonous plants in the family Euphorbiaceae. Adults often migrate, in search of better food for their larvae.

• OCCURRENCE Tropical and subtropical regions. On host plants.

• REMARK South American and Madagascan species are so large and colourful that they are often mistaken for butterflies.

CATERPILLARS of some species live together in a silk web when young.

CHRYSIRIDIA RIPHEARIA, the Madagascan Sunset Moth, has yellow-and-black caterpillars that are distasteful to predators.

forewings have pointed tip

distinctive spotting and three tails on hindwings

Wingspan 6–10cm (2½–4in)	Larval feeding habits

Order LEPIDOPTERA	Family ZYGAENIDAE	No. of species 800

BURNET MOTHS

Also called foresters, many of these moths are black with bright or metallic red, green, or blue colouring. Some look wasp-like. The antennae are thickened, and the head has a pair of hairy protuberances above the eyes. Most species produce hydrogen cyanide – a feature advertised by warning coloration.

• LIFE-CYCLE Eggs are laid on herbaceous host plants, and larvae feed externally on their leaves. The adults fly and feed at flowers by day. Pupation takes place within a long cocoon.

• OCCURRENCE Worldwide, except New Zealand; especially tropics and subtropics. On host plants.

CATERPILLARS have small warts, from which tufts of hairs protrude.

slender body

△ *ADSCITA STATICES*, the Forester Moth, is found in Europe and Asia. The caterpillars feed on sorrel.

black head

black body

bold coloration indicates distastefulness

ZYGAENA FILIPENDULAE, also known as the Six-spot Burnet, is common throughout Europe.

Wingspan 2.5–3.5cm (1–1¼in)	Larval feeding habits

Order LEPIDOPTERA	Family LYCAENIDAE	No. of species 6,000

BLUES, COPPERS, AND HAIRSTREAKS

The males and females of these small, slender-bodied butterflies are often differently coloured. The upper wings of males can be iridescent blue, coppery, or purplish but some are brown or orange. The undersides of both sexes are dull, with small, dark-centred spots.

• LIFE-CYCLE Eggs are laid on host plants. The caterpillars feed on plants or on aphids, coccids, and other small insects. The larvae of many species secrete a special fluid that is eaten by ants. In return, the ants guard them from enemies and allow them to eat the ants' larvae. Pupation occurs on the host plant, in debris, or underground.

• OCCURRENCE Worldwide, especially in warmer regions. In association with ants' nests or with host plants.

upper wing surfaces bright and iridescent

many species have tails on their hindwings

△ *THECLA CORONATA*, the Hewitson's Blue Hairstreak, is found in tropical regions of South America. This is one of the largest and most brilliantly coloured members of the family.

CATERPILLARS are slug-like, with a squat, tapering shape, and are green or brown.

wings of male are vivid violet-blue

black markings on forewings

orange edging to hindwings

△ *POLYOMMATUS ICARUS*, known as the Common Blue, is one of the most widespread European natives.

white scales on body resemble hairs

△ *LYCAENA PHLAEAS*, the Small Copper, is a colourful and very common little butterfly, found across the northern hemisphere.

highly distinctive orange markings on forewings of female

white streaks on wings

short, dark-edged tails

UNDERSIDE

THECLA BETULAE, the Brown Hairstreak Butterfly, is a woodland species found in Europe and temperate regions of Asia.

Wingspan 1.5–5cm (⅝–2in)	Larval feeding habits

Order LEPIDOPTERA	Family NYMPHALIDAE	No. of species 5,000

NYMPHALIDS

Also called brush-foots, due to the
greatly reduced front pair of legs,
species of this huge family fly by
day and vary enormously in size and
colour. The upper surfaces of the
wings are usually brightly coloured,
but the undersides have camouflage
colouring to protect the butterfly at rest.
• LIFE-CYCLE Females lay groups of
round, ribbed eggs on the foliage of trees,
shrubs, and herbaceous plants. Caterpillars
may feed communally when very young.
Nymphalid pupae, which often have warty
bumps, hang head down from host plants by
a small group of terminal hooks (a cremaster).
• OCCURRENCE Worldwide. Usually in
flower-filled meadows and woodland clearings.
• REMARK A very small number of
nymphalid species can be pests, causing
damage to crops such as sweet
potato and soyabean.

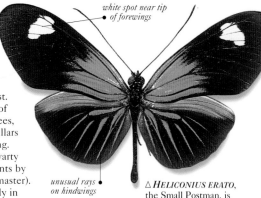

white spot near tip
of forewings

unusual rays
on hindwings

△ *HELICONIUS ERATO*,
the Small Postman, is
native to Central and
South America.

CATERPILLARS are
generally spiny with
branched bumps and
projections.

▷ *NYMPHALIS ANTIOPA* is
often called the Camberwell
Beauty, referring to the part of
south London where the first
recorded specimen was found.

pale spots on margin
of forewings

distinctive
row of purple-
blue spots

distinctive white
spots on forewings

broad red
bands

red, black-spotted
hindwings

VANESSA ATALANTA, or the Red
Admiral, is a widespread species.
The caterpillars, usually black
with white and yellow spots, feed
on nettles and related plants.

Wingspan 3–15cm (1¼–6in)	Larval feeding habits

▽ *MORPHO MENELAUS* lives in South America. The female is shown here. Males are almost entirely metallic blue.

"toothed" white markings on wing margins

scalloped wing margins

white patches all around wing margins

white-centred eye-spots

▽ **UPPERSIDE**

white patches on head

▽ **UNDERSIDE**

scalloped hindwing margins

DANAUS PLEXIPPUS, or the Monarch Butterfly, is famous for its long-distance migrations, the longest of which extends from Mexico to Canada.

MANIOLA JURTINA, the Meadow Brown, is a very common species in the pastures and meadows of Europe. Its caterpillars feed on grasses.

Order LEPIDOPTERA	Family PAPILIONIDAE	No. of species 600

SWALLOWTAILS

The wings of these butterflies are typically dark, with
bands, spots, or patches of
white, yellow, orange, red,
green, or blue. True to their
name, many species have
tails on their hindwings.
• LIFE-CYCLE Round eggs are
laid on host plants – the caterpillars eat
a range of foliage. Pupation usually occurs
on the host plant with the chrysalis
upright and held on by a silk "belt".
• OCCURRENCE Worldwide,
especially in warmer regions. In flower-
rich habitats, either open or shaded.
• REMARK This family contains the
birdwing species – the world's largest
butterflies, now protected by law.

scalloped wing margins

CATERPILLARS
emit an odour –
from a thoracic
scent gland – that
deters predators.

tails on hindwings

PAPILIO GLAUCUS, the
Tiger Swallowtail, is a North
American species. The
common name refers to its
forewing markings.

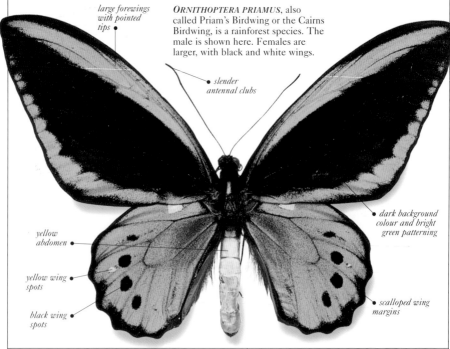

large forewings with pointed tips

ORNITHOPTERA PRIAMUS, also
called Priam's Birdwing or the Cairns
Birdwing, is a rainforest species. The
male is shown here. Females are
larger, with black and white wings.

slender antennal clubs

dark background colour and bright green patterning

yellow abdomen

yellow wing spots

black wing spots

scalloped wing margins

Wingspan 4.5–28cm (1¾–11in)	Larval feeding habits

Order LEPIDOPTERA	Family PIERIDAE	No. of species 1,200

WHITES AND SULPHURS

The wings of these very common butterflies are usually white, yellow, or orange with black or dark grey markings. The pigmentation of the wing scales comes from by-products of food eaten by the caterpillars.

• **LIFE-CYCLE** Females typically lay single, elongate, ribbed eggs on a wide range of host plants. The caterpillars have no spines or projections, but the pupae have a distinctive spiny projection that arises from the head end, and are held upright on the host plant by a silk "belt".

• **OCCURRENCE** Worldwide. In a wide range of habitats. They are often seen in groups around bird droppings, urine, or puddles in sunshine. Some species migrate in large numbers.

• **REMARK** Many species are crop pests. The Large White (*Pieris brassicae*) and the Small White (*P. rapae*) are serious pests of cabbage crops.

distinctive orange tip on male's forewings •

△ *ANTHOCHARIS CARDAMINES* occurs across Europe and Asia. In both sexes, the undersides of the hindwings are mottled green. The male has orange tips on its forewings.

CATERPILLARS may have camouflage colouring – green with pale stripes or bands. Many have spine-like hairs.

grey tip on • forewings

dark mark • on pale wings

hindwing has • whitish upperside and pale yellow underside with scattered grey scales

△ *PIERIS RAPAE*, the Small White, is an unremarkable looking species that is found all over the world. The male is shown here; the female is yellowish in colour.

black spot on forewings •

broad, dark wing margins •

COLIAS EURYTHEME, the Orange Sulphur, is widespread in North America. The male, shown here, is slightly smaller than the female.

• orange spot on hindwings

Wingspan 2–7cm (¾–2¾in)	Larval feeding habits

BEES, WASPS, ANTS, AND SAWFLIES

T HE ORDER Hymenoptera contains 91 families and 198,000 species. These are further divided into two suborders: primitive, plant-eating insects called sawflies (Symphyta) and wasps, ants, and bees (Apocrita). The families shown here are arranged in three groups: social wasps and bees and ants (families Andrenidae to Vespidae), parasitic wasps (Agaonidae to Trichogrammatidae), and sawflies (Argidae to Siricidae).

Most members of the order have two pairs of membranous wings, joined in flight by tiny hooks. In all species except the sawflies, the first abdominal segment is fused to the thorax, while the second and sometimes the third segments are narrow and form a waist. Sawfly females have a saw-like ovipositor while female parasitic wasps often have a long, slender ovipositor, which may also be internal. The ovipositor of female bees, ants, and social wasps has evolved into a sting. Eggs issue from an opening at its base. Metamorphosis is complete. Gender is determined by haplodiploidy, a process in which fertilized eggs produce females and males arise from unfertilized eggs.

Many species show advanced forms of social behaviour and play a vital role in various types of ecosystem as predators, parasites, and plant pollinators.

Order HYMENOPTERA	Family ANDRENIDAE	No. of species 2,500

MINING BEES

These honeybee-like species are typically red-brown or brown-black, although some are yellow or white. The thorax and abdomen may be hairy. Most species are solitary.
• **LIFE-CYCLE** Females make nests in soil burrows and lay eggs in specially prepared cells, where the larvae develop. These cells are usually coated with a protective waterproof substance secreted by an abdominal gland and are supplied with pollen and honey – food for the larvae. Pollen is collected and taken back to the nest on the bees' hindlegs.
• **OCCURRENCE** Worldwide, except Australia. In flower-rich habitats, especially during spring.
• **REMARK** Mining bees are vital pollinators of spring flowers.

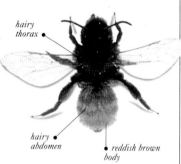

hairy thorax

hairy abdomen

reddish brown body

ANDRENA FULVA, the Tawny Mining Bee, appears in early spring. It often nests in lawns, making a little mound of earth at the opening.

white hairs

ANDRENA CINERARIA is a European bee. The female is shown here. The male has a hairier thorax, and white hairs on all femora. Females have white hair only on the femora of the front legs.

no white hairs in centre of thorax

abdomen dark with bluish sheen

LARVAE can be quite slender or stout. There are protruberances on the abdominal segments.

Length 0.4–2cm (⁵⁄₃₂–¾in), most 1–1.5cm (⅜–⅝in)	Larval feeding habits ✳

Order HYMENOPTERA	Family ANTHOPHORIDAE	No. of species 4,000

CUCKOO, DIGGER, AND CARPENTER BEES

Cuckoo bees are often wasp-like and are black and yellow or brown and white. They do not have a pollen-carrying region on their hindlegs. Digger bees are generally stout and hairy, while carpenter bees are either very large and black or bluish or small and dark blue-green. Most species are solitary.

• **LIFE-CYCLE** Cuckoo bees parasitize the nests of soil-nesting bees. They lay their eggs in these nests, and the cuckoo bee larva then hatches out, kills the occupying egg or larva, and feeds on its provisions. Digger bees make their nests in ground burrows and supply the nest's larval cells with pollen and honey. Carpenter bees dig tunnels in solid wood, in which they prepare brood cells. Each cell is supplied with a mass of sticky pollen, on which the female lays a single large egg before sealing the cell with chewed wood fibres.

• **OCCURRENCE** Worldwide. In a wide variety of flower-rich habitats.

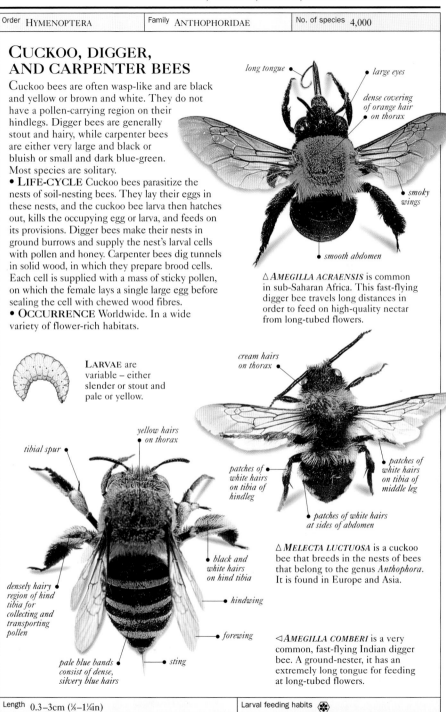

long tongue

large eyes

dense covering of orange hair on thorax

smoky wings

smooth abdomen

△ *AMEGILLA ACRAENSIS* is common in sub-Saharan Africa. This fast-flying digger bee travels long distances in order to feed on high-quality nectar from long-tubed flowers.

LARVAE are variable – either slender or stout and pale or yellow.

cream hairs on thorax

yellow hairs on thorax

tibial spur

patches of white hairs on tibia of hindleg

patches of white hairs on tibia of middle leg

patches of white hairs at sides of abdomen

black and white hairs on hind tibia

△ *MELECTA LUCTUOSA* is a cuckoo bee that breeds in the nests of bees that belong to the genus *Anthophora*. It is found in Europe and Asia.

densely hairy region of hind tibia for collecting and transporting pollen

hindwing

forewing

pale blue bands consist of dense, silvery blue hairs

sting

◁ *AMEGILLA COMBERI* is a very common, fast-flying Indian digger bee. A ground-nester, it has an extremely long tongue for feeding at long-tubed flowers.

Length 0.3–3cm (⅛–1¼in)	Larval feeding habits ✿

Order HYMENOPTERA	Family APIDAE	No. of species 1,000

HONEYBEES AND THEIR RELATIVES

The most familiar members of this family are the stout, very hairy bumblebees and the smaller, more slender honeybees. Most females have a special pollen basket (corbiculum) on the outside of the hind tibiae. Coloration is highly varied.

• LIFE-CYCLE These bees are social and live in colonies consisting of an egg-laying queen, males (drones), and sterile worker females who find food and look after the young. Bumblebees form small colonies under or on the ground. The nests in which they lay their eggs are made of grass with wax brood cells. Honeybee colonies comprise a queen, up to 2,000 males, and thousands of workers. The nest is an array of double-sided wax combs divided into hexagonal cells for rearing young and storing pollen and honey. Workers use a dance language to convey the distance, quality, and direction of food.

• OCCURRENCE Worldwide, except in sub-Saharan Africa. Bumblebees are very common in northern temperate regions. In well-vegetated, flower-rich habitats.

• REMARK As well as providing honey, wax, and other products, these bees pollinate most of the world's plants.

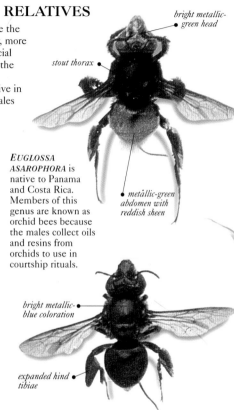

bright metallic-green head

stout thorax

metallic-green abdomen with reddish sheen

EUGLOSSA ASAROPHORA is native to Panama and Costa Rica. Members of this genus are known as orchid bees because the males collect oils and resins from orchids to use in courtship rituals.

bright metallic-blue coloration

expanded hind tibiae

LARVAE are pale and grub-like. Bumblebee larvae are fatter than honeybee larvae.

△ EUGLOSSA INTERSECTA is native to Surinam, Guyana, and parts of northern Brazil. Like most Euglossa species, it has bright, metallic coloration.

• hexagonal cells of honeycomb, made of wax

• workers tending larvae

◁ APIS MELLIFERA, the Western Honeybee, is now found worldwide and is the best known member of the honeybee genus Apis. Millions of trips between flowers and the hive are required to make one jar of honey.

Length 0.3–3cm (⅛–1¼in)	Larval feeding habits

▷ *PSITHYRUS* SPECIES are closely associated with bumblebees. They are cuckoo bees that lay their eggs in the nests of *Bombus* species. Many *Psithyrus* bees closely resemble bumblebees, especially the species that they parasitize.

wings often darker tinged than those of bumblebees

abdomen less hairy than that of bumblebee

pollen basket absent

hairy hind tibiae

pale yellow "collar"

fringe of pale hairs at rear margin of thorax

mostly red abdomen

◁*BOMBUS MONTICOLA* is a relatively small, distinctive bumblebee. As its name implies, it is common in upland and mountainous areas. It makes a nest in underground burrows, often close to bilberry plants.

compound eye

yellow-orange "collar"

pollen load carried on hindleg

lemon-yellow "collar"

hairy hind tibiae

△ *BOMBUS LUCORUM* is an extremely common bumblebee species. It makes its nest under the ground and is one of the first bees to be seen in early summer in Europe.

△ *BOMBUS TERRESTRIS* is known as the Buff-tailed Bumblebee because the workers and males always have a whitish abdominal tail. Here, a sterile female worker bee feeds at a flower.

Order HYMENOPTERA	Family BETHYLIDAE	No. of species 2,000

BETHYLIDS

Usually black or brownish, these wasps have quite elongate heads. Some females look ant-like; certain others have a pitted surface and look similar to velvet ants (see p.187). Wings may be present in both sexes but females are often wingless.
• **LIFE-CYCLE** Females lay eggs on the outside of hosts such as beetle larvae or moth caterpillars – either a host she has found in a sheltered spot, or one that she has paralysed and dragged to such a place. She may stay with the larvae as they develop.
• **OCCURRENCE** Worldwide, especially in warm regions. In varied habitats, where hosts are found.

BETHYLUS SPECIES have very potent stings for their small size. Hosts are either paralysed or killed.

• *large head compared with thorax*

• *stout femora*

• *black coloration*

LARVAE are pale and fatter towards the rear.

Length 0.4–2cm (⁵⁄₃₂–³⁄₄in), most under 1cm (³⁄₈in)	Larval feeding habits

Order HYMENOPTERA	Family CHRYSIDIDAE	No. of species 3,000

JEWEL WASPS

Also called cuckoo wasps, because they steal their host larvae's provisions, or ruby-tailed wasps, due to their colouring, most species of jewel wasps are bright metallic blue, green, red, or combinations of these colours. Their hard, dimpled body protects them from bee and wasp stings.
• **LIFE-CYCLE** Typically, the female finds a nest containing the larva of a solitary bee or wasp and lays an egg. The jewel wasp larva eats the host larva from the outside, plus the host's provisions.
• **OCCURRENCE** Worldwide. In a variety of habitats, wherever their hosts are found.

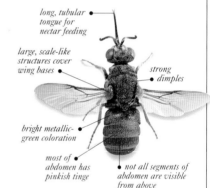

long, tubular tongue for nectar feeding •

large, scale-like structures cover wing bases •

• *strong dimples*

bright metallic-green coloration •

most of abdomen has pinkish tinge •

• *not all segments of abdomen are visible from above*

△ *PARNOPES CARNEA* is a European species. Members of this genus parasitize solitary hunting wasps.

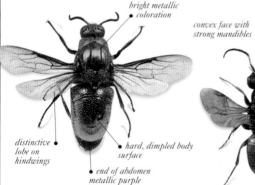

bright metallic coloration •

distinctive lobe on hindwings •

• *hard, dimpled body surface*

• *end of abdomen metallic purple*

convex face with strong mandibles •

• *bright metallic-purple body*

• *dark femora*

• *orange abdomen with greenish tinge at end*

LARVAE are smooth and stout. The middle of the body is the broadest part.

STILBUM SPLENDIDUM is a large species native to northern Australia. It parasitizes solitary mud-nesting wasps.

***CLEPTES* SPECIES** are found mostly in the northern hemisphere. The body is often not entirely metallic.

Length 0.2–2cm (¹⁄₁₆–³⁄₄in), most under 1.2cm (¹⁄₂in)	Larval feeding habits

| Order HYMENOPTERA | Family COLLETIDAE | No. of species 2,000 |

PLASTERER AND YELLOW-FACED BEES

These solitary bees are slender to fairly robust, and most are very dark or black. The body hairs are pale golden or white, and the abdominal hairs often form bands.
• **LIFE-CYCLE** Plasterer bees dig burrows in soil, waterproofing the cells with a special abdominal secretion. Yellow-faced bees nest in hollow plant stems and the burrows of wood-boring insects. Each larval cell is supplied with regurgitated pollen and nectar.
• **OCCURRENCE** Worldwide, especially in the southern hemisphere. Common at flowers.
• **REMARK** Plasterer bees carry pollen on their hindlegs, while yellow-faced bees carry it in a special pouch known as a crop.

heart-shaped face

smooth body

black-and-orange banded abdomen

△ *HYLAEOIDES CONCINNA* is a yellow-faced bee with a distinct, but reddish, facial mark.

conspicuous fringes of white hairs on abdominal segments

long white hairs on legs

LARVAE are variable but are generally curved and maggot-like.

◁ *COLLETES DAVIESANUS* is a plasterer bee, native to Europe. It makes its nest in the vertical faces of sandy cliffs.

| Length 0.3–1.8cm (⅛–¾in), most under 1.3cm (½in) | Larval feeding habits |

| Order HYMENOPTERA | Family DRYINIDAE | No. of species 1,000 |

DRYINID WASPS

The males of these mainly brown or black wasps are winged. Females may be wingless or ant-like, and their front tarsus often forms a kind of claw.
• **LIFE-CYCLE** Females hunt down the nymphs or adults of certain bugs, sting them, and lay an egg inside them. The wasp larva feeds on the host's fluids, developing in a larval sac that protrudes from the host's body and then emerges from this to pupate inside a cocoon.
• **OCCURRENCE** Worldwide. In various habitats, wherever hosts are found.

GONATOPUS SEPSOIDES is a British wasp. Like many members of this genus, the female looks very ant-like.

large head relative to thorax

front femora elongate and swollen in middle

claw-like tarsal segment for holding hosts down

pronotum broader than long

ant-like body

dark pterostigma

shiny black body

LARVAE are usually pale, have large heads, and are strongly curved or U-shaped.

◁ *CHELOGYNUS SCAPULARIS* is a European wasp. Its distinctive legs are yellowish with much darker femora.

| Length 0.2–1.2cm (1/16–½in), most under 0.8cm (5/16in) | Larval feeding habits |

Order HYMENOPTERA	Family FORMICIDAE	No. of species 9,000

ANTS

These highly social insects live in colonies that consist of a dozen to several million individuals. The most commonly seen ants are the sterile, wingless, female workers. Reproductive queens and males usually have wings. The second, or second and third, segments of the abdomen are constricted to form a distinct "waist". This waist may have either bumps or spiny processes. Most ant species are red-brown to black in colour, but yellows and greens also occur. Ants protect themselves by biting or stinging, or by spraying formic acid.

• **LIFE-CYCLE** After mating, males die and queens shed their wings. Typically, a single queen lays all of a colony's eggs. As the colony grows, workers take away and protect the eggs, and then care for and feed the hatched young. If a protein diet is fed to female larvae, they become reproductives.

• **OCCURRENCE** Worldwide. In all habitats.

• **REMARK** Ants are significant predators or herbivores in most habitats. Much more animal flesh, for example, is eaten by ants in African savannahs than by lions, hyenas, and other carnivores. Some species, such as the leaf-cutter ants (*Atta* species) and the Fire Ant (*Solenopsis invicta*), can be serious crop pests.

ants pulling leaves together   *larva*   *worker using larva to "sew" edges of leaf together*

OECOPHYLLA SMARAGDINA, the Green Tree or Weaver Ant, builds shelters out of leaves. Workers pull the edges of leaves together while other workers use silk produced by the larvae to stick the edges together. The larva is held in the worker's mandibles and used like a shuttle.

LARVAE are white, grub-like, and slightly curved. There may be hairs on the body.

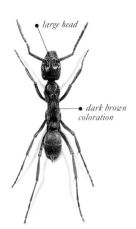

large head

dark brown coloration

MEGAPONERA FOETENS is found in Africa and is a predator of termites. Once a termite has been found, the ants lay trails of pheromones back to the nest to recruit more workers.

▽ *MYRMECIA* SPECIES, the Australian bulldog ants, can be fairly large. Workers have a powerful sting.

mandibles

very narrow "waist"

rounded abdomen

long antennae

toothed mandibles

head of some Dinoponera ants may be wider than 4mm (⅛in)

DINOPONERA GRANDIS is native to parts of South America. The large workers are solitary hunters, and the colonies are small.

Length 0.1–2cm (½–¾in)	Larval feeding habits

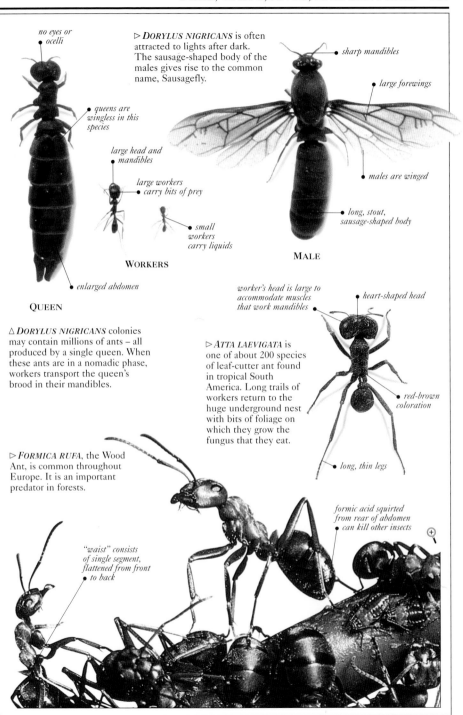

no eyes or ocelli

▷ **DORYLUS NIGRICANS** is often attracted to lights after dark. The sausage-shaped body of the males gives rise to the common name, Sausagefly.

sharp mandibles

queens are wingless in this species

large forewings

large head and mandibles

large workers carry bits of prey

males are winged

small workers carry liquids

long, stout, sausage-shaped body

WORKERS

MALE

enlarged abdomen

QUEEN

△ **DORYLUS NIGRICANS** colonies may contain millions of ants – all produced by a single queen. When these ants are in a nomadic phase, workers transport the queen's brood in their mandibles.

worker's head is large to accommodate muscles that work mandibles

heart-shaped head

▷ **ATTA LAEVIGATA** is one of about 200 species of leaf-cutter ant found in tropical South America. Long trails of workers return to the huge underground nest with bits of foliage on which they grow the fungus that they eat.

red-brown coloration

▷ **FORMICA RUFA**, the Wood Ant, is common throughout Europe. It is an important predator in forests.

long, thin legs

formic acid squirted from rear of abdomen can kill other insects

"waist" consists of single segment, flattened from front to back

Order HYMENOPTERA	Family HALICTIDAE	No. of species 3,500

SWEAT BEES

Despite the name, only a few species in this family are attracted to sweat. Most are brown or black, but some have a metallic-blue or green sheen. The body may be pitted or dimpled with only a sparse covering of hairs. Many species are solitary, while others are social to various degrees.

• **LIFE-CYCLE** Eggs are laid inside nests made in soil or in rotten wood. The cells in which eggs are brooded are waterproofed with a secretion that sticks to the surrounding soil and prevents fungal growth.

• **OCCURRENCE** Worldwide. Widespread, especially in flower-rich areas and woodland margins.

single groove
under socket of
each antenna

dark brown
body with
yellow bands

very
hairy legs

LARVAE may have bumps on the upper surface, as well as tiny spines.

HALICTUS QUADRICINCTUS is found across southern Europe and the Mediterranean. It is one of the largest European species in its genus.

Length 4–15mm (⅛–⅝in), most under 10mm (⅜in)	Larval feeding habits

Order HYMENOPTERA	Family MEGACHILIDAE	No. of species 3,000

LEAF-CUTTER AND MASON BEES

Most of these bees are solitary. Many have stout, dark brown to black bodies and may have yellow or pale markings; some are metallic blue or green. Pollen-collecting species carry their loads in a brush of hairs found underneath the abdomen.

• **LIFE-CYCLE** Most species lay eggs in nests made in the natural cavities of dead wood, hollow stems, and snail shells. Leaf-cutter bees cut circular pieces of leaves or petals to line the nests' brood cells. Other species use hairs from woolly-leaved plants. Mason bees make mud cells under stones and in burrows. Some species use the nests of other bees rather than making their own.

• **OCCURRENCE** Worldwide. In a variety of habitats.

• **REMARK** Vital crop-pollinators, these bees may be taken from crop to crop by farmers on huge trailers.

CHALICODOMA MONTICOLA nests in hollow plant stems, such as bamboo canes, and builds cell partitions out of mud or from a mud and resin mixture.

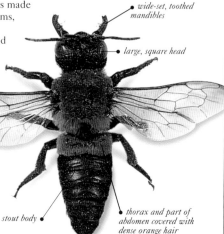

wide-set, toothed
mandibles

large, square head

stout body

thorax and part of
abdomen covered with
dense orange hair

LARVAE are stout, and are often fatter towards the rear end of the body.

Length 0.7–2cm (¼–¾in)	Larval feeding habits

Order HYMENOPTERA	Family MUTILLIDAE	No. of species 5,000

VELVET ANTS

These wasps are referred to as velvet ants because the females are covered with soft, velvety hairs and are wingless and ant-like. The males have fully developed wings. Velvet ants are black or red-brown, with spots or bands of short hairs that are red, yellow, or silver. The body surface has coarse dimples.

• **LIFE-CYCLE** Velvet ants use the larvae and pupae of other wasps and bees – those that make soil, wood, or paper nests – as food for their larvae. On finding a suitable host brood cell, the female bites it open. She will re-seal any cell where the larva inside is too young. If there is a fully grown larva or pre-pupa inside, however, she will lay an egg on it, before re-sealing the cell. The hatched velvet ant larva then eats the host larva and pupates inside the cell.

• **OCCURRENCE** Worldwide, especially in subtropical and tropical regions. Females are often seen on the ground in dry habitats.

• **REMARK** Female velvet ants have very powerful stings.

LARVAE have very rounded abdominal segments (on the upperside) when seen in profile.

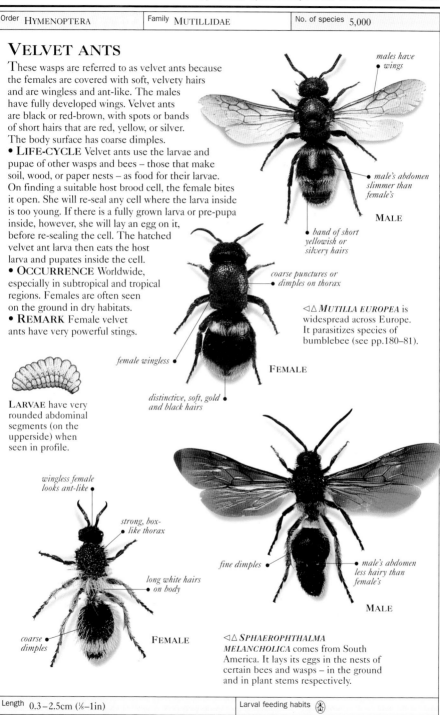

males have wings

male's abdomen slimmer than female's

MALE

band of short yellowish or silvery hairs

coarse punctures or dimples on thorax

◁△ *MUTILLA EUROPEA* is widespread across Europe. It parasitizes species of bumblebee (see pp.180–81).

female wingless

FEMALE

distinctive, soft, gold and black hairs

wingless female looks ant-like

strong, box-like thorax

long white hairs on body

coarse dimples

FEMALE

fine dimples

male's abdomen less hairy than female's

MALE

◁△ *SPHAEROPHTHALMA MELANCHOLICA* comes from South America. It lays its eggs in the nests of certain bees and wasps – in the ground and in plant stems respectively.

Length 0.3–2.5cm (⅛–1in)	Larval feeding habits

Order HYMENOPTERA	Family POMPILIDAE	No. of species 4,000

SPIDER-HUNTING WASPS

Most spider-hunting wasps are dark blue or black
with wings in shades of dark yellow, blue, or black.
The body is slender, and the hindlegs are long and
spiny. Males are smaller and more slender than the
females. Some species can be very large.

• LIFE-CYCLE Females fly or run along the
ground in search of spiders. The wasps have to
wrestle with their prey, but their strong venom
can cripple even very large specimens. The female
drags the paralysed spider to a prepared mud nest
in a crevice or under the ground, although some
will attack a spider in its own burrow. Before
sealing the nest, the female lays a single egg,
usually on the spider's abdomen. Some lay their
eggs on a spider caught by another wasp before
it is sealed in; others open already sealed nests.

• OCCURRENCE Worldwide, especially in
tropical and subtropical regions. In varied habitats,
where spiders are found.

• REMARK The stings
of these wasps are often
extremely painful.

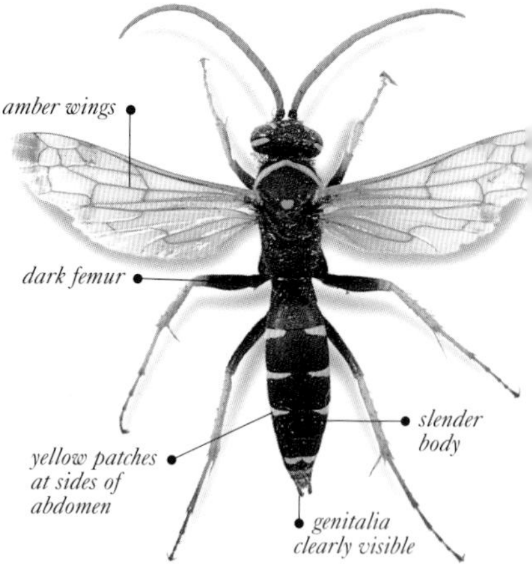

amber wings

dark femur

*yellow patches
at sides of
abdomen*

*slender
body*

*genitalia
clearly visible*

△ *POMPILUS* SPECIES
males (seen here) have
12 antennal segments,
while females have 13.
There are six visible
segments on the male
abdomen; seven on
that of the female.

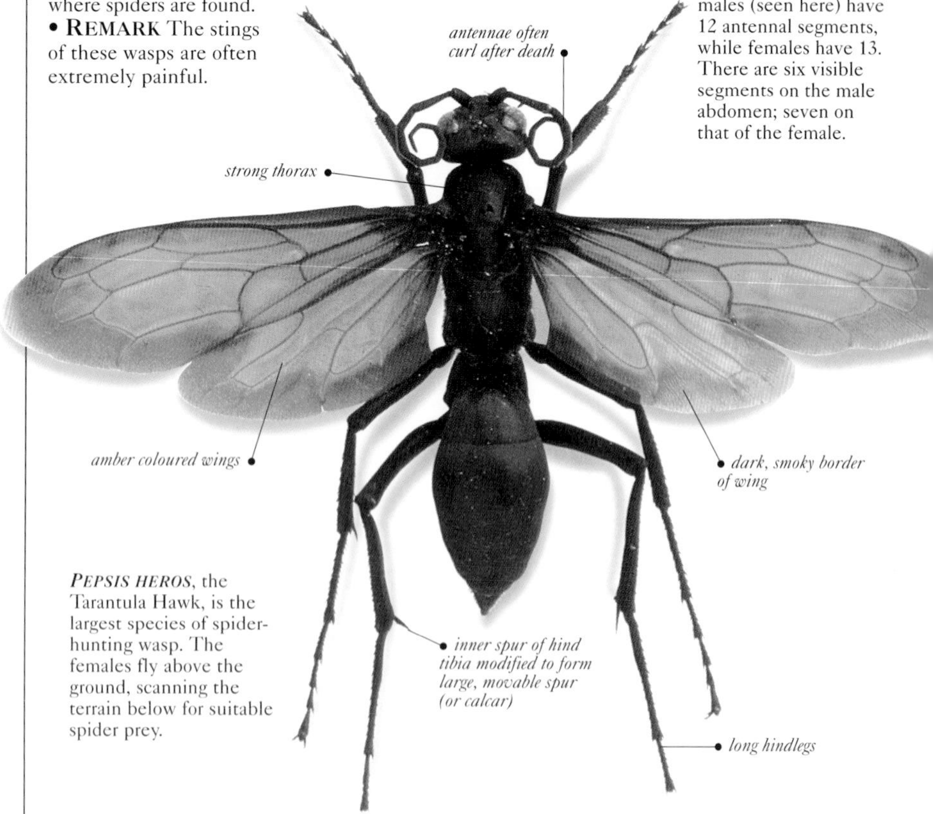

*antennae often
curl after death*

strong thorax

amber coloured wings

*dark, smoky border
of wing*

PEPSIS HEROS, the
Tarantula Hawk, is the
largest species of spider-
hunting wasp. The
females fly above the
ground, scanning the
terrain below for suitable
spider prey.

*inner spur of hind
tibia modified to form
large, movable spur
(or calcar)*

long hindlegs

Length 0.5–7cm (³⁄₁₆–2¾in), most under 2.5cm (1in)	Larval feeding habits 🕷

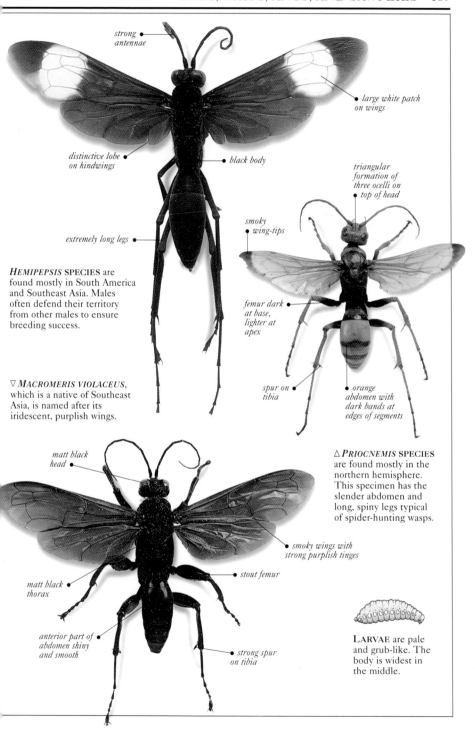

strong antennae

large white patch on wings

distinctive lobe on hindwings

black body

triangular formation of three ocelli on top of head

smoky wing-tips

extremely long legs

HEMIPEPSIS SPECIES are found mostly in South America and Southeast Asia. Males often defend their territory from other males to ensure breeding success.

femur dark at base, lighter at apex

∇ **MACROMERIS VIOLACEUS**, which is a native of Southeast Asia, is named after its iridescent, purplish wings.

spur on tibia

orange abdomen with dark bands at edges of segments

matt black head

△ **PRIOCNEMIS SPECIES** are found mostly in the northern hemisphere. This specimen has the slender abdomen and long, spiny legs typical of spider-hunting wasps.

smoky wings with strong purplish tinges

stout femur

matt black thorax

anterior part of abdomen shiny and smooth

strong spur on tibia

LARVAE are pale and grub-like. The body is widest in the middle.

Order HYMENOPTERA	Family SCOLIIDAE	No. of species 350

MAMMOTH WASPS

True to their name, these are very large
wasps, with stout bodies. They have dark
coloration: bluish black with reddish brown
markings. The ends of the wings, which may
be clear, smoky, metallic blue, or
orangish, appear finely wrinkled.
The body is densely covered
with dark or gold-coloured hair.
Male mammoth wasps are smaller
and slimmer than the females and have
longer, thicker antennae. In both sexes, there is a
noticeable notch on the inside margin of the eye.
• LIFE-CYCLE After mating, females hunt
through leaf-litter or dig under the ground in search
of the larvae of scarab beetles to use as hosts. Once
found, the female wasp stings and paralyses a beetle
larva and lays a single egg on the outside. When the
wasp larva emerges, it consumes the beetle larva
and pupates inside a tough cocoon that it spins
alongside the host's remains.
• OCCURRENCE Worldwide, mainly in tropical
regions. In various habitats, where scarab hosts occur.

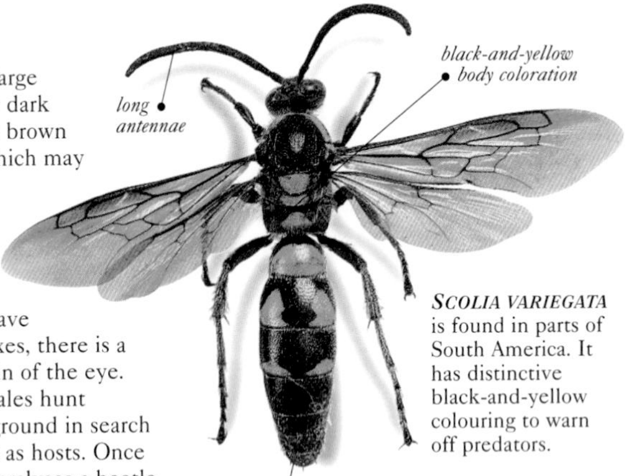

long antennae

black-and-yellow body coloration

SCOLIA VARIEGATA
is found in parts of
South America. It
has distinctive
black-and-yellow
colouring to warn
off predators.

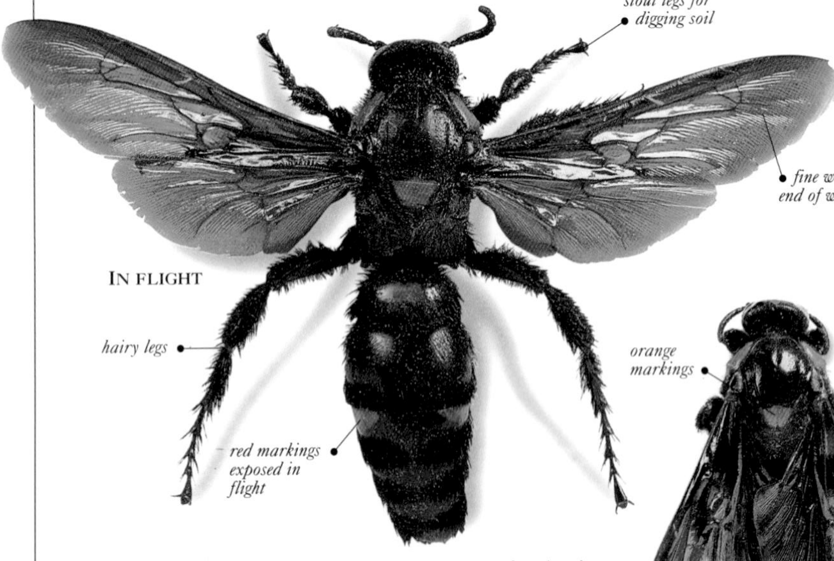

males have
three short
spines at end
of abdomen

stout legs for
digging soil

fine wrinkles at
end of wings

IN FLIGHT

hairy legs

red markings
exposed in
flight

orange
markings

forewings larger
than hindwings

AT REST

dense black
hairs

SCOLIA PROCER comes from Java,
Borneo, and Sumatra. Members
of this genus are not especially
aggressive but may sting painfully
if handled carelessly.

Length 1–5.6cm (⅜–2¼in)	Larval feeding habits 🦂

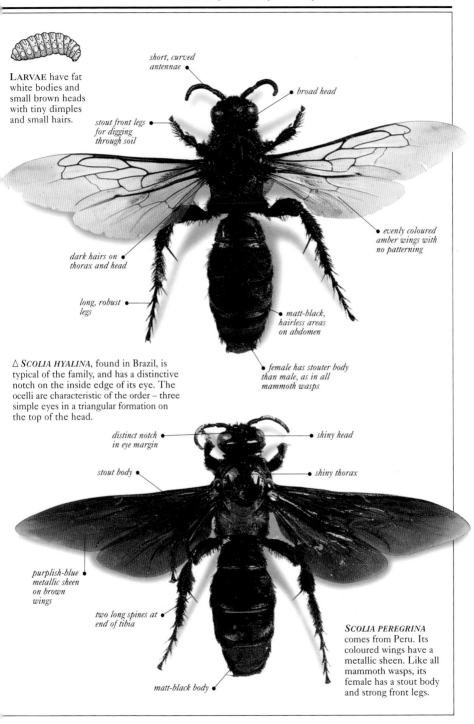

LARVAE have fat white bodies and small brown heads with tiny dimples and small hairs.

short, curved antennae

broad head

stout front legs for digging through soil

dark hairs on thorax and head

long, robust legs

evenly coloured amber wings with no patterning

matt-black, hairless areas on abdomen

female has stouter body than male, as in all mammoth wasps

△ *SCOLIA HYALINA*, found in Brazil, is typical of the family, and has a distinctive notch on the inside edge of its eye. The ocelli are characteristic of the order – three simple eyes in a triangular formation on the top of the head.

distinct notch in eye margin

shiny head

stout body

shiny thorax

purplish-blue metallic sheen on brown wings

two long spines at end of tibia

matt-black body

SCOLIA PEREGRINA comes from Peru. Its coloured wings have a metallic sheen. Like all mammoth wasps, its female has a stout body and strong front legs.

Order HYMENOPTERA	Family SPHECIDAE	No. of species 8,000

DIGGER WASPS

Some digger species are known as solitary hunting wasps, sand wasps, or mud dauber wasps. Relatively hairless and often brightly coloured, they are found in many forms. They are all solitary and nest in plant stems, soil, or rotten wood. Some nest inside insect burrows. Not all species actually dig a nest.
• LIFE-CYCLE Typically, a female catches an insect or a spider, paralyses it, and takes it to a prepared nest, where it is buried – along with a wasp egg – for the emerging larva to eat. There may be one or several prey items, or prey may be added as the larvae grow. A few species lay eggs in the nest of other digger wasps.
• OCCURRENCE Worldwide. In a range of habitats.

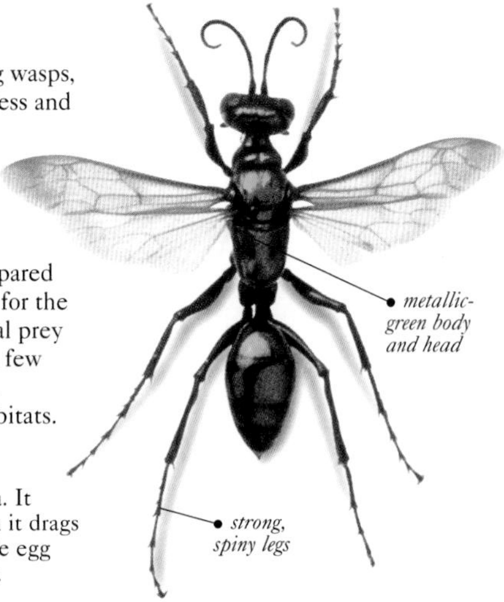

metallic-green body and head

strong, spiny legs

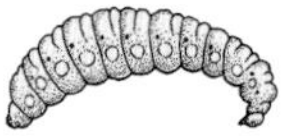

▷CHLORION LOBATUM is found in Southeast Asia. It parasitizes crickets, which it drags to a burrow. It lays a single egg before sealing the burrow.

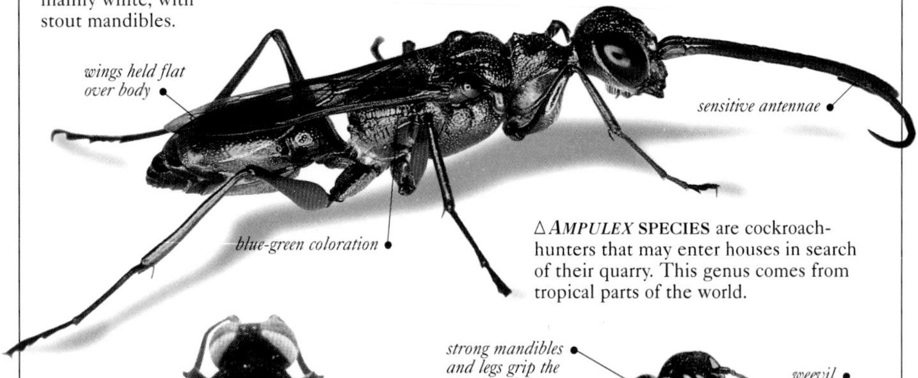

LARVAE taper at both ends and are mainly white, with stout mandibles.

wings held flat over body

sensitive antennae

blue-green coloration

△*AMPULEX* SPECIES are cockroach-hunters that may enter houses in search of their quarry. This genus comes from tropical parts of the world.

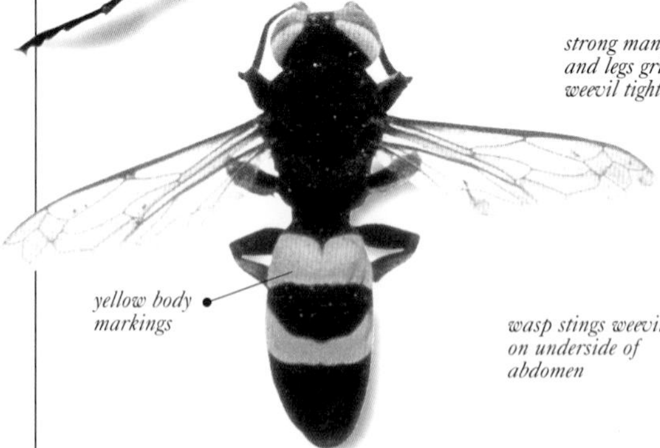

strong mandibles and legs grip the weevil tightly

weevil

yellow body markings

wasp stings weevil on underside of abdomen

EDITHA MAGNIFICA, from Brazil, is a very brightly marked insect that specializes in catching butterflies, mainly of the family Pieridae (see p.177)

CERCERIS ARENARIA, the Weevil-hunting Wasp, catches weevils and paralyses them with its sting. The weevils are buried in wasp burrows, in which the wasp then lays its eggs.

Length 0.4–4.8cm (5⁄32–2in)	Larval feeding habits 🐜 ✺

Order HYMENOPTERA	Family TIPHIIDAE	No. of species 1,600

TIPHIID WASPS

Tiphiids are shiny and vary from quite slender to stout. Some species have wingless females that look like ants.

• **LIFE-CYCLE** During mating, the female may be carried aloft by the male. She then hunts out hosts for her larvae – usually larvae of scarab, longhorn, or tiger beetles. After paralysing a host with her sting, she lays an egg on its body.

• **OCCURRENCE** Worldwide. In a range of habitats.

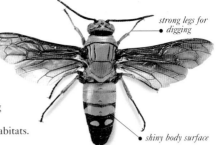

strong legs for digging

shiny body surface

LARVAE are long and white, with bumps on the upper abdominal surface.

THYNNUS VENTRALIS comes from Australia. Its females are wingless, and its larvae parasitize scarab-beetle larvae (see p.128).

Length 0.5–2.8cm (³⁄₁₆–1¼in)	Larval feeding habits

Order HYMENOPTERA	Family VESPIDAE	No. of species 4,000

SOCIAL WASPS

These wasps roll or fold their wings longitudinally, rather than hold them flat over the body. Nearly all have warning coloration in shades of brown or black and orange or yellow. The most familiar species are paper wasps and yellow jackets, whose nests are made of chewed fibres. However, the family also includes mason wasps, which make mud-lined, underground nests in stems and crevices.

• **LIFE-CYCLE** Paper wasps and yellow jackets have queens and workers, co-operate in caring for the brood, and have overlapping generations. A typical queen overwinters, makes a nest in spring, and rears the first small brood herself. Inside the nest, larvae develop within the cells of horizontal "combs" and are fed chewed-up insects by sterile female workers. As the colony grows, so does the nest.

• **OCCURRENCE** Worldwide. In a wide range of habitats.

• **REMARK** Like several insects in the order Hymenoptera, a social wasp can deliver a painful sting.

LARVAE have bodies that are widest about a third back from the head.

orange abdomen, with dark markings

△ ***VESPA CRABRO***, the Hornet, makes its nest in hollow trees. Its colonies tend to consist of only a few hundred workers.

wings folded longitudinally

VESPULA VULGARIS is very helpful in the garden as it removes caterpillars and other pests. It makes its nest from wood fibres.

▷ ***VESPULA GERMANICA*** is found in warmer parts of the world. Colonies of this species may be perennial, with more than one queen.

yellow patches on thorax

black dots on abdomen

Length 0.4–3.6cm (³⁄₁₆–1¼in)	Larval feeding habits

Order HYMENOPTERA	Family AGAONIDAE	No. of species 650

FIG WASPS

The males and females of this family look very different. The tiny, flat-bodied females have wings. Males hardly resemble wasps at all – most are wingless with odd-shaped heads, weak middle legs, and their abdomen folded underneath their body. The common name of this family is derived from the fact that these wasps and fig trees are totally dependent on each other. The trees can be pollinated only by these wasps, which in turn are able to reproduce only inside figs. Each wasp species pollinates a particular fig species.

• **LIFE-CYCLE** Life-cycles can be complex. Typically, a female enters a young fig through a hole. The inside of the fig is lined with female flowers and the wasp pollinates these and lays eggs in some of the ovules. The larvae develop here and feed on galls produced during the egg-laying process. Males usually emerge first and mate with the females before they emerge, biting through the female's gall wall to reach her. By this time, the male flowers inside the fig have produced pollen, which the departing females pick up and take to the next fig tree.

• **OCCURRENCE** Worldwide, in tropical, subtropical, and warm temperate regions. Wherever fig trees grow.

• **REMARK** Some species are parasites, laying eggs inside the larvae of other pollinating fig wasps. Recent research, however, puts some of these parasites in other families.

LARVAE are small, pale, and grub-like. They develop inside figs.

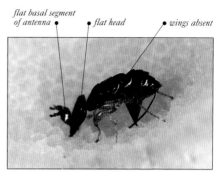

flat basal segment of antenna • *flat head* • *wings absent*

CERATOSOLEN MEGACEPHALUS is an African ⊕ fig wasp. The female is shown here, in the process of laying her eggs inside a fig. She has lost her wings and the ends of her antennae in the struggle to enter the fig.

strong front legs • *shiny black body* • *ovipositor*

BLASTOPHAGA PSENES is found all over the ⊕ world. It pollinates *Ficus carica*, the common fig. A female is shown here, sitting on the outside of a fig. There are ten times more females than males.

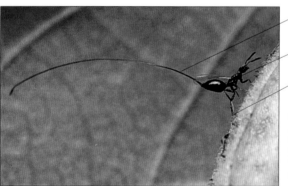

• *extremely long ovipositor*

• *metallic green-blue colouring*

• *pale tarsi*

SYCOSCAPTER **SPECIES** are parasitic fig wasps, found in Africa. Recent DNA analysis suggests that these wasps might actually belong to the family Pteromalidae (see p.201).

Length 1–3mm (¹⁄₃₂–¹⁄₈in)	Larval feeding habits

Order HYMENOPTERA	Family BRACONIDAE	No. of species 25,000

BRACONID WASPS

Most species of braconid wasp are
small and inconspicuous, and brown,
reddish-brown, or black in colour. Some
have very faint patterns of veins on their wings.
• **LIFE-CYCLE** Braconid wasps are parasites, and
each species uses a different host – mainly butterfly
and moth caterpillars, but the young of aphids, flies,
or other insects are used by some. A few species
are hyperparasitoids. Females
lay eggs on or inside the
host. If the host is large,
there may be enough food
for hundreds of wasp larvae to
develop inside. Larvae that
develop inside aphids, which are
found on foliage, spin a silk cocoon that
sticks the host's body to the leaf. These
mummified aphids contain the pupating wasp.
The emerging adult wasp cuts a neat hole in the
corpse and flies off.
• **OCCURRENCE** Worldwide. In a variety
of habitats, wherever suitable hosts are found.
• **REMARK** Many species are used to control
populations of insect pests.

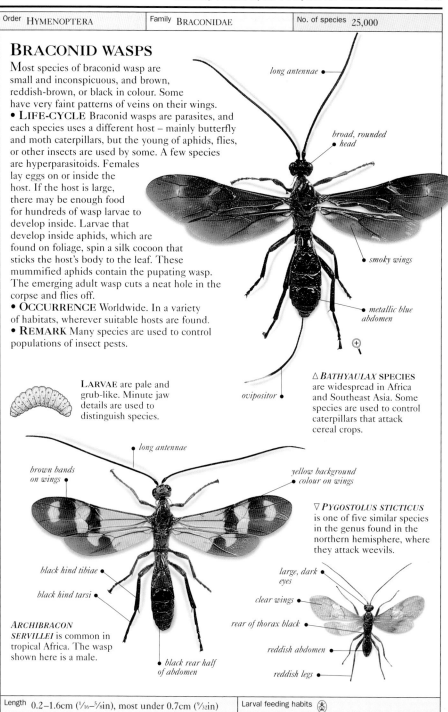

long antennae

broad, rounded head

smoky wings

metallic blue abdomen

ovipositor

LARVAE are pale and
grub-like. Minute jaw
details are used to
distinguish species.

△ *BATHYAULAX* SPECIES
are widespread in Africa
and Southeast Asia. Some
species are used to control
caterpillars that attack
cereal crops.

long antennae

brown bands on wings

yellow background colour on wings

▽ *PYGOSTOLUS STICTICUS*
is one of five similar species
in the genus found in the
northern hemisphere, where
they attack weevils.

black hind tibiae

black hind tarsi

large, dark eyes

clear wings

rear of thorax black

reddish abdomen

reddish legs

*ARCHIBRACON
SERVILLEI* is common in
tropical Africa. The wasp
shown here is a male.

black rear half of abdomen

Length 0.2–1.6cm (1/16–5/8in), most under 0.7cm (9/32in)	Larval feeding habits

Order HYMENOPTERA	Family CHALCIDIDAE	No. of species 1,800

CHALCID WASPS

Most chalcid wasps are dark brown, black, red, or yellow. The body may have sculpturing or pits and occasionally has a metallic sheen. The first hindleg segment is large, and the hind femora are greatly enlarged and toothed underneath. Females have a short, inconspicuous ovipositor.
• LIFE-CYCLE Eggs are laid inside the larvae and pupae of other insects. Some species are hyperparasitoids.
• OCCURRENCE Worldwide. In various habitats, wherever suitable hosts are found.

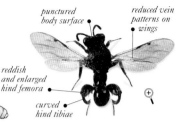

punctured body surface

reduced vein patterns on wings

reddish and enlarged hind femora

curved hind tibiae

LARVAE are white and grub-like and have small heads.

CHALCIS SISPES is native to parts of Europe and Asia. Its larvae parasitize the larvae of soldier flies (see p.152).

Length 0.2–1.5cm (¹⁄₁₆–⁵⁄₈in), most under 0.8cm (⁵⁄₁₆in)	Larval feeding habits

Order HYMENOPTERA	Family CYNIPIDAE	No. of species 1,250

GALL WASPS

These wasps are shiny red-brown or black and usually have fully developed wings. The thorax has a humped appearance, and the abdomen of the female is flat.
• LIFE-CYCLE The female lays her eggs inside the tissue of oak species or other woody plants. This induces the host plant to develop a swollen gall that protects and nourishes the developing larvae. Galls vary enormously in size, colour, texture, and location and may contain one or more developing larvae.
• OCCURRENCE Worldwide, mostly in the northern hemisphere. In a variety of habitats, wherever suitable host trees and plants grow.
• REMARK The tissue inside plant galls may support diverse communities of organisms, many of which are parasitic wasps.

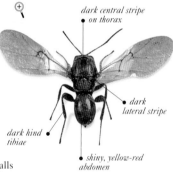

dark central stripe on thorax

dark lateral stripe

dark hind tibiae

shiny, yellow-red abdomen

ANDRICUS QUERCUSRADICIS is a widespread European wasp that uses many different oak species as its hosts.

shiny, punctured surface

large thorax

shiny, smooth abdomen

◁**ANDRICUS SPECIES** are widespread in Europe. Many are very similar to each other in appearance.

LARVAE are pale, grub-like, smooth, and often taper towards the rear.

Length 1–9mm (¹⁄₃₂–¹¹⁄₃₂in)	Larval feeding habits

Order HYMENOPTERA	Family ENCYRTIDAE	No. of species 3,800

ENCYRTID WASPS

This large family is quite variable, especially in the appearance of the head and antennae. Most of these small species are robust, slender, or slightly flat. They can be orange, red, or brown, often with a metallic sheen. The thorax is convex, and the middle legs, which are used for jumping, have a large, curved tibial spur.
• LIFE-CYCLE The females of most species locate and lay eggs in the nymphs and adults of other insects – generally scale insects, mealybugs, aphids, and whiteflies. Some, however, specialize in parasitizing caterpillars or weevil grubs. A few are hyperparasitoids. Sometimes the eggs divide repeatedly to produce anything from 10 to 2,000 larvae, depending on the size of the host. Pupation occurs inside the host's body.
• OCCURRENCE Worldwide. In a wide variety of habitats, wherever hosts are found.
• REMARK These wasps are among the most important biological control agents, and many species have been used against serious crop pests. *Copidosoma koehleri*, for example, is used to control the Potato Tuber Moth in India.

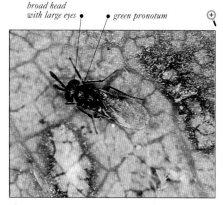

broad head with large eyes • *green pronotum*

COPIDOSOMA SPECIES is found in parts of Europe and Asia. Their hosts are various moth species, including those belonging to the family Noctuidae (see p.165).

LARVAE vary, but may be pale and taper evenly towards the rear.

Length 0.5–4.5mm (¹⁄₆₄–⁵⁄₃₂in), most 1–2mm (¹⁄₃₂–¹⁄₁₆in)	Larval feeding habits

Order HYMENOPTERA	Family EULOPHIDAE	No. of species 3,400

EULOPHID WASPS

These small wasps vary from elongate to stout. The body is soft, and the antennae have fewer than ten segments. They may be yellow, brown, or black, sometimes with a metallic sheen.
• LIFE-CYCLE The females of most species hunt for the larvae of leaf-miners and gall-formers in which to lay their eggs, but some attack the larvae or pupae of moths, beetles, flies, and bugs. Certain small species even use the eggs of insects as hosts, and a few are hyperparasitoids. The developing wasp larva consumes the host and pupates.
• OCCURRENCE Worldwide. In a wide variety of habitats, wherever hosts are found.
• REMARK Eulophid wasps destroy various insect pests, and many are used for specific biological control programmes.

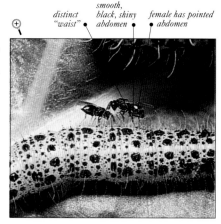

distinct "waist" • *smooth, black, shiny abdomen* • *female has pointed abdomen*

TETRASTICUS GALACTOPUS is a hyperparasitoid. It attacks the larvae of the parasitic wasp *Cotesia glomeratus*, which itself is found inside the body of its own host – the Cabbage White Butterfly.

LARVAE are often pale and grub-like with small heads and stout bodies.

Length 0.5–5mm (¹⁄₆₄–³⁄₁₆in), most 1–3mm (¹⁄₃₂–¹⁄₈in)	Larval feeding habits

Order HYMENOPTERA	Family EURYTOMIDAE	No. of species 1,400

EURYTOMID WASPS

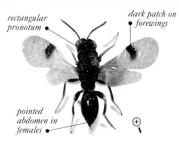

rectangular pronotum

dark patch on forewings

pointed abdomen in females

These wasps are yellow, reddish, or dull black. A few have a metallic sheen. They look similar to chalcid wasps (see p.196), but the hind coxae are never very enlarged and the femora do not have projections.
• **LIFE-CYCLE** Many of these wasps lay eggs inside seeds, where their larvae develop. Some are leaf-miners or gall-formers. Others develop as parasitoids inside beetle, wasp, or fly larvae, and the smaller species attack the eggs of grasshoppers or certain bugs. A few species have a mixed feeding strategy, parasitizing gall-forming insects initially and then, as the larvae grow bigger, eating the gall tissue.
• **OCCURRENCE** Worldwide. In a variety of habitats.

SYCOPHILA BIGUTTATA develops inside galls made by gall wasps on oak trees. The larvae of this species are parasitic on the gall-former.

dimpled thorax

ovipositor placed inside gall tissue to reach gall wasp larvae

EURYTOMA BRUNNIVENTRIS is linked with certain gall-forming wasps. Its larvae may parasitize the wasps or other insects inside the gall. They may also eat gall tissue.

LARVAE are tiny, white, and grub-like. Some have quite long hairs.

Length 2–6mm (¹⁄₁₆–¹⁄₄in)		Larval feeding habits

Order HYMENOPTERA	Family GASTERUPTIIDAE	No. of species 500

GASTERUPTIID WASPS

These slender, dark-coloured wasps look very like ichneumons (see opposite), but the head is carried on a short neck, and the slim abdomen joins the thorax well above the hind coxae. The hindlegs are long, and the hind tibia are swollen at their ends. The ovipositor can be very long.
• **LIFE-CYCLE** After mating, females seek out the nests of solitary bees or wasps in sandy soil or inside plant stems or wood. Eggs are laid in the nest, and the larvae eat the eggs and the food store left for the host bee larvae.
• **OCCURRENCE** Worldwide, especially in warmer areas. In various habitats, wherever hosts are found.
• **REMARK** Gasteruptiids have a distinctive hovering flight, with their hindlegs dangling below the body.

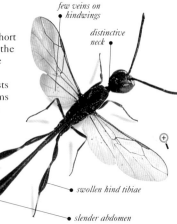

few veins on hindwings

distinctive neck

swollen hind tibiae

slender abdomen

LARVAE can be quite hairy, and the mandibles have three teeth.

very long, slender ovipositor

GASTERUPTION SPECIES are often seen feeding at flowers in the sunshine.

Length 1.2–2.8cm (¹⁄₂–1¹⁄₄in)		Larval feeding habits

Order HYMENOPTERA	Family ICHNEUMONIDAE	No. of species 60,000

ICHNEUMON WASPS

These generally slender wasps may be yellowish brown to black, and may have either brown and black or yellow and black patterning. The slender abdomen is usually joined to the thorax by a thin stalk. The ovipositor is typically long and clearly visible, although it is short in some species.

• **LIFE-CYCLE**
Females mainly attack the larvae and pupae of insects such as beetles, flies, moths, sawflies, and other wasps. Some species use spiders as hosts. They use their long ovipositors to lay eggs on or inside the host.

• **OCCURRENCE** Worldwide, especially in temperate areas. In a wide range of habitats where hosts occur.

• **REMARK** Many species benefit humans by controlling populations of other insects.

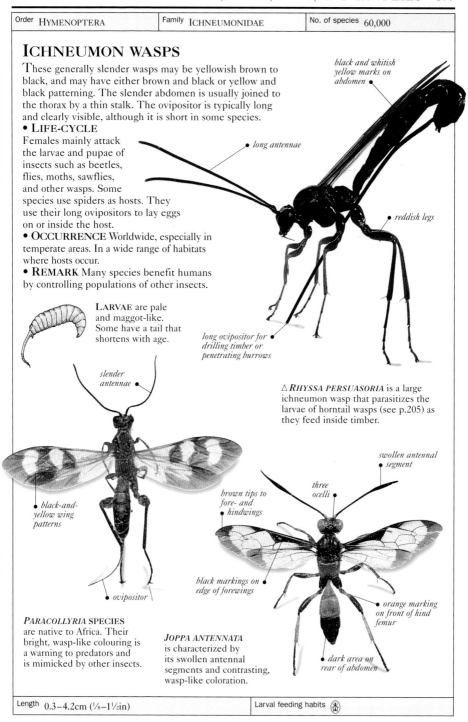

black and whitish yellow marks on abdomen

long antennae

reddish legs

LARVAE are pale and maggot-like. Some have a tail that shortens with age.

long ovipositor for drilling timber or penetrating burrows

slender antennae

black-and-yellow wing patterns

ovipositor

△ *RHYSSA PERSUASORIA* is a large ichneumon wasp that parasitizes the larvae of horntail wasps (see p.205) as they feed inside timber.

swollen antennal segment

three ocelli

brown tips to fore- and hindwings

black markings on edge of forewings

orange marking on front of hind femur

dark area on rear of abdomen

PARACOLLYRIA SPECIES are native to Africa. Their bright, wasp-like colouring is a warning to predators and is mimicked by other insects.

JOPPA ANTENNATA is characterized by its swollen antennal segments and contrasting, wasp-like coloration.

Length 0.3–4.2cm (⅛–1½in)		Larval feeding habits 🐛

Order HYMENOPTERA	Family MYMARIDAE	No. of species 1,400

FAIRYFLIES

This family includes the world's smallest
flying insects. They are dark brown, black, or
yellow in coloration, but are never metallic. The
narrow forewings lack any conspicuous vein
pattern but have a distinctive fringe of hairs.
The stalked and strap-like hindwings are
also fringed with minute hairs.
• **LIFE-CYCLE** The females of all species
parasitize the eggs of other insects. Most
specialize on the eggs of plant-hoppers and
other bug families, but the eggs of a range
of other insects are also
used as hosts.
• **OCCURRENCE**
Worldwide. In a wide
variety of habitats,
wherever hosts are found.
• **REMARK** Several
species have been used
to control insect pests.

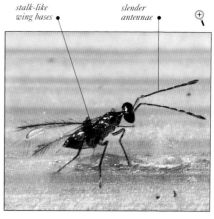

stalk-like
wing bases •

slender
antennae •

LARVAE are tiny
and tailed at first,
and grub-like at a
later stage.

ANAGRUS OPTABILIS is a specialist parasitoid
of the eggs of certain plant-hoppers (the family
Delphacidae). Related species have been used
to control plant-hoppers that attack rice crops.

Length 0.2–5mm (¹⁄₂₈–³⁄₁₆in), most 0.5–1.5mm (¹⁄₆₄–¹⁄₁₆in)	Larval feeding habits ⊛

Order HYMENOPTERA	Family PROCTOTRUPIDAE	No. of species 500

PROCTOTRUPIDS

Most of the species in this family are either very
dark or black in colour and smooth-surfaced. The
abdomen tapers at both ends and is often paler
than the thorax and head. There is a conspicuous
pterostigma on the relatively large forewings.
• **LIFE-CYCLE** Females seek out the larvae of
beetles, and sometimes of gall midges, that live in
leaf-litter or decaying wood and lay eggs inside
them. When fully grown, the larva chews a hole
through the membrane between two abdominal
segments of its host and emerges almost
completely. It pupates with its
rear end still in contact with
the host's remains.
• **OCCURRENCE**
Worldwide. In
woodland and a
range of moist
habitats.

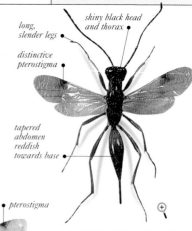

long,
slender legs •

shiny black head
and thorax •

distinctive
pterostigma •

tapered
abdomen
reddish
towards base •

• pterostigma

△ *PROCTOTRUPES GRAVIDATOR*
is found throughout the northern
hemisphere and in parts of
Southeast Asia. It parasitizes
ground-beetle larvae (see p.112).

LARVAE are
small, smooth,
pale, and
grub-like.

EXALLONYX LONGICORNIS
is widespread in Europe and
Asia, where it parasitizes rove-
beetle larvae (see p.130).

Length 0.3–1cm (⅛–⅜in), most under 0.8cm (⁵⁄₁₆in)	Larval feeding habits ⊛

Order HYMENOPTERA	Family PTEROMALIDAE	No. of species 4,000

PTEROMALID WASPS

Most of these slim to quite robust wasps are black, metallic-blue, metallic-green, or green- or yellow-brown. The thorax is often dimpled. Viewed from the side, the smooth abdomen is frequently triangular in females and oblong in males.

• LIFE-CYCLE Pteromalids have quite varied life-cycles. The larvae may be endo- or ectoparasitoids or hyperparasitoids. Most species use the larvae or pupae of flies, beetles, wasps, fleas, butterflies, and moths as hosts. Females may have to drill through plant tissue to reach gall-forming, leaf-mining, or stem-boring hosts. Some species lay just a single egg, but others lay hundreds of eggs if the host is large enough.

• OCCURRENCE Worldwide. In a wide variety of habitats, wherever hosts are found.

• REMARK Some species are used to control populations of harmful crop pests.

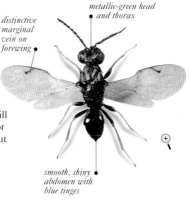

distinctive marginal vein on forewing

metallic-green head and thorax

smooth, shiny abdomen with blue tinges

LARVAE are pale and grub-like, with a small head. Some have small bumps on the upper or lower body surface.

PTEROMALUS SPECIES are common parasitoids. Their larvae develop inside the larvae and pupae of a wide range of insects.

Length 1–8mm (¹⁄₃₂–⁵⁄₁₆in), most under 5mm (³⁄₁₆in)	Larval feeding habits

Order HYMENOPTERA	Family SCELIONIDAE	No. of species 3,000

SCELIONID WASPS

These wasps are typically black, although they may be yellow or brown. The body shape varies from quite slender to quite robust, and the abdomen is generally flat, with sharply angled side margins.

• LIFE-CYCLE The females of most species lay eggs in the eggs of other insects, especially those belonging to the orders Lepidoptera, Hemiptera, Coleoptera, and Orthoptera. Some species hang on to a host insect until it lays its eggs. To prevent another wasp laying eggs in an egg that they have parasitized, females mark a host's egg with an odour. The hatched scelionid larva feeds on the tissues of the host's eggs and pupates inside.

• OCCURRENCE Worldwide. Widespread in many habitats, but especially common in open grassland. Some are specialist parasites of mantids and grasshoppers in semi-arid areas and deserts.

• REMARK These wasps are parasitic on some crop pests, and several species have been used in pest-control programmes.

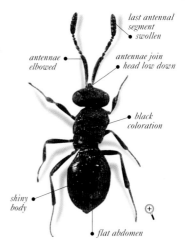

last antennal segment swollen

antennae elbowed

antennae join head low down

black coloration

shiny body

flat abdomen

TRIMORUS PEDESTRE is found in Europe and Asia. Both sexes are wingless. This species has no reason to fly because it parasitizes the eggs of ground-living beetles.

LARVAE are pale and grub-like, with a flat rear. The head is often withdrawn into the thorax.

Length 0.05–1cm (¹⁄₆₄–³⁄₈in), most under 3mm (¹⁄₈in)	Larval feeding habits

Order HYMENOPTERA	Family TORYMIDAE	No. of species 1,250

TORYMID WASPS

These wasps are usually elongate in shape, with metallic blue or green coloration. The thorax has dimples, while the abdomen is smooth.

• LIFE-CYCLE Most species parasitize gall-forming flies and gall-wasps. In some cases, females use their ovipositors to drill through gall tissue and lay eggs on the host larvae inside. Other species parasitize caterpillars, mantid egg cases, and the larvae of some bees and wasps. Herbivorous species produce larvae that develop in the seeds of various trees.

• OCCURRENCE Worldwide. In a wide range of habitats, wherever suitable hosts can be found.

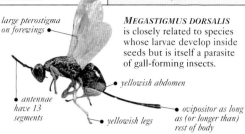

large pterostigma on forewings •

MEGASTIGMUS DORSALIS is closely related to species whose larvae develop inside seeds but is itself a parasite of gall-forming insects.

• antennae have 13 segments

• yellowish abdomen

• yellowish legs

• ovipositor as long as (or longer than) rest of body

▽ **TORYMUS SPECIES** include many wasps that seek out larvae inside gall tissue. The ovipositor is often very long, to reach into large galls.

antennae can sense vibrations from deep within gall •

shiny, metallic, • dimpled surface

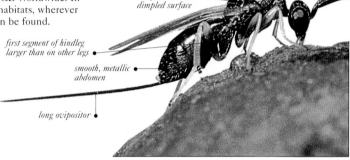

first segment of hindleg larger than on other legs •

smooth, metallic • abdomen

long ovipositor •

LARVAE are pale, grub-like, and often hairy.

Length 0.1–1.4cm (½₂–⅗in), most under 0.5cm (³⁄₁₆in)	Larval feeding habits

Order HYMENOPTERA	Family TRICHOGRAMMATIDAE	No. of species 600

TRICHOGRAMMATID WASPS

Because they are so small, these wasps are often overlooked. Most species are pale and fairly stout bodied. The veinless wings have small hairs forming distinctive lines across the surface and a fringe around the edge.

• LIFE-CYCLE Eggs are laid inside the eggs of many other insects. Larval development and pupation can take as little as three days.

• OCCURRENCE Worldwide. In a wide range of habitats, anywhere that insect eggs can be found – usually exposed, on foliage.

eggs being parasitized •

broad head •

black, shiny thorax •

LARVAE are pale, featureless, minute grubs, found inside host eggs.

TRICHOGRAMMA SEMBLIDIS, like other related species, has been used to control many butterfly and moth pests worldwide. Here, the eggs being parasitized are those of the alderfly (*Sialis lutaria*).

Length 0.3–1.2mm (¹⁄₂₈–¹⁄₂₂in)	Larval feeding habits

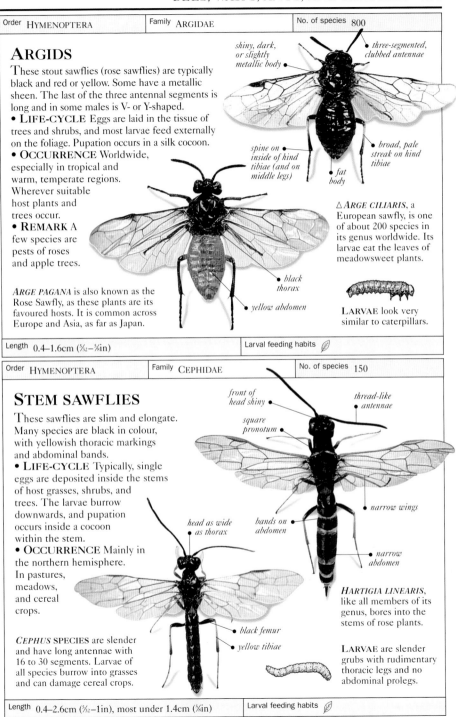

| Order HYMENOPTERA | Family ARGIDAE | No. of species 800 |

ARGIDS

These stout sawflies (rose sawflies) are typically black and red or yellow. Some have a metallic sheen. The last of the three antennal segments is long and in some males is V- or Y-shaped.
• **LIFE-CYCLE** Eggs are laid in the tissue of trees and shrubs, and most larvae feed externally on the foliage. Pupation occurs in a silk cocoon.
• **OCCURRENCE** Worldwide, especially in tropical and warm, temperate regions. Wherever suitable host plants and trees occur.
• **REMARK** A few species are pests of roses and apple trees.

shiny, dark, or slightly metallic body

three-segmented, clubbed antennae

spine on inside of hind tibiae (and on middle legs)

broad, pale streak on hind tibiae

fat body

△ *ARGE CILIARIS*, a European sawfly, is one of about 200 species in its genus worldwide. Its larvae eat the leaves of meadowsweet plants.

ARGE PAGANA is also known as the Rose Sawfly, as these plants are its favoured hosts. It is common across Europe and Asia, as far as Japan.

black thorax

yellow abdomen

LARVAE look very similar to caterpillars.

| Length 0.4–1.6cm (⁵⁄₃₂–⅝in) | Larval feeding habits 🍃 |

| Order HYMENOPTERA | Family CEPHIDAE | No. of species 150 |

STEM SAWFLIES

These sawflies are slim and elongate. Many species are black in colour, with yellowish thoracic markings and abdominal bands.
• **LIFE-CYCLE** Typically, single eggs are deposited inside the stems of host grasses, shrubs, and trees. The larvae burrow downwards, and pupation occurs inside a cocoon within the stem.
• **OCCURRENCE** Mainly in the northern hemisphere. In pastures, meadows, and cereal crops.

front of head shiny

thread-like antennae

square pronotum

narrow wings

head as wide as thorax

bands on abdomen

narrow abdomen

HARTIGIA LINEARIS, like all members of its genus, bores into the stems of rose plants.

CEPHUS **SPECIES** are slender and have long antennae with 16 to 30 segments. Larvae of all species burrow into grasses and can damage cereal crops.

black femur

yellow tibiae

LARVAE are slender grubs with rudimentary thoracic legs and no abdominal prolegs.

| Length 0.4–2.6cm (⁵⁄₃₂–1in), most under 1.4cm (⅝in) | Larval feeding habits 🍃 |

Order HYMENOPTERA	Family CIMBICIDAE	No. of species 150

CIMBICID SAWFLIES

These large sawflies may resemble hairless bees.
Many are fat-bodied, with slightly flat abdomens. Most
are black or yellowish and black. The antennae have fewer
than seven segments, and the last one is swollen.
• LIFE-CYCLE Eggs are laid in host plants. Larvae feed
externally on foliage, and pupation takes place in a cocoon.
• OCCURRENCE Northern
hemisphere, South
America, and eastern
Asia. Wherever host
plants are found.

*slightly
flattened
abdomen*

*clubbed
antennae*

△ *PACHYLOSTICA VIOLA*
is found only in South
America, where its larvae
probably eat tree foliage.

CIMBEX FEMORATUS is a   *fat body*
heavy-bodied British species
whose larvae feed on birch trees.

LARVAE are slightly
curved, with thoracic legs
and abdominal prolegs.

Length 2–3cm (¾–1¼in)	Larval feeding habits

Order HYMENOPTERA	Family PAMPHILIIDAE	No. of species 200

LEAF-ROLLING SAWFLIES

These sawflies have a broad head and a flat
body. Most species are strongly built and are
black with yellow markings.
• LIFE-CYCLE Eggs are laid in plant tissue.
The larvae feed under silk webs or in foliage,
singly or in groups.
• OCCURRENCE
Northern hemisphere.
In trees and shrubs in
a variety of habitats.

flat body

broad head

*ACANTHOLYDA
POSTICALIS* was
once found only in
Scottish pine woods
but is now found in
conifer plantations
further south.

LARVAE do not have
abdominal prolegs.

Length 1–1.5cm (⅜–⅝in)	Larval feeding habits

Order HYMENOPTERA	Family PERGIDAE	No. of species 350

PERGID SAWFLIES

These fairly robust species have a
rounded abdomen. The antennae
may be simple, branched, or sawlike.
• LIFE-CYCLE Eggs are laid in
plant tissue. The larvae typically
feed in groups on foliage. They can
produce chemicals to repel predators.
• OCCURRENCE Southern
hemisphere, except Africa. Wherever
host trees, especially eucalyptus, occur.
• REMARK Several species in
Australia are pests in eucalyptus
forests and can cause total defoliation.

PERGA DORSALIS is typical
of Australian eucalyptus-
feeding species whose
larvae can be serious pests
of these trees.

*antennae end in
small clubs*

*orange
marks*

*dark
leading edge
of forewing*

LARVAE have short
prolegs and may have
warning coloration.

stout thorax

Length 0.4–2.5cm (³⁄₁₆–1in)	Larval feeding habits

Order HYMENOPTERA	Family SIRICIDAE	No. of species 100

HORNTAILS

Also known as wood wasps, these large sawflies are reddish brown, black and yellow, or metallic blue-purple. The common name refers to a spine at the end of the abdomen.

• LIFE-CYCLE The females drill into the wood of live or fallen trees to lay one egg at a time and infect the tree with a rot-producing fungus. The larvae burrow into heartwood and eat the fungus and the wood. Pupation occurs in a cocoon of silk and chewed wood.

• OCCURRENCE Worldwide. In coniferous and deciduous woodland.

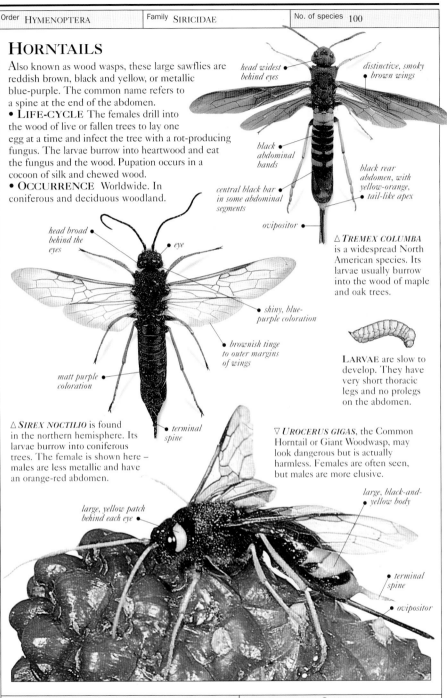

head widest • behind eyes

distinctive, smoky • brown wings

black abdominal bands

central black bar • in some abdominal segments

ovipositor •

black rear abdomen, with yellow-orange, • tail-like apex

△ **TREMEX COLUMBA** is a widespread North American species. Its larvae usually burrow into the wood of maple and oak trees.

head broad • behind the eyes

• eye

• shiny, blue-purple coloration

• brownish tinge to outer margins of wings

matt purple • coloration

• terminal spine

△ **SIREX NOCTILIO** is found in the northern hemisphere. Its larvae burrow into coniferous trees. The female is shown here – males are less metallic and have an orange-red abdomen.

LARVAE are slow to develop. They have very short thoracic legs and no prolegs on the abdomen.

▽ **UROCERUS GIGAS**, the Common Horntail or Giant Woodwasp, may look dangerous but is actually harmless. Females are often seen, but males are more elusive.

large, black-and-• yellow body

large, yellow patch behind each eye •

• terminal spine

• ovipositor

Length 2–4cm (¾–1⅛in)	Larval feeding habits

Order HYMENOPTERA	Family TENTHREDINIDAE	No. of species 6,000

COMMON SAWFLIES

These sawflies are highly variable in both their appearance and habits. They may be brown, black, or green, brightly coloured, or even wasp-like. The sexes may be differently coloured.

• LIFE-CYCLE Females use their ovipositor to cut slits in the leaves, twigs, and shoots of host plants and lay eggs inside. The larvae feed on the outside of the plants and are solitary with camouflage colouring or gregarious with warning coloration. Some mine leaves or make galls. Pupation occurs inside a silk cocoon, either under the ground or in leaf-litter.

• OCCURRENCE Worldwide, except in New Zealand, especially in northern and cool temperate regions. In gardens, pastures, and woodland.

• REMARK Many species are pests, causing serious damage to fruit and vegetable crops and forest trees.

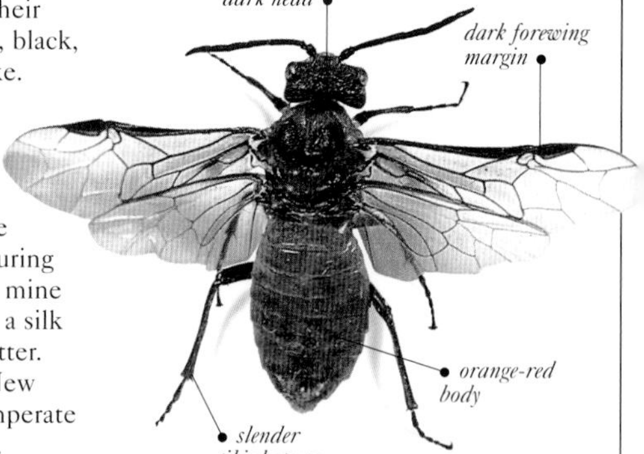

dark head •

dark forewing margin •

• orange-red body

• slender tibial spurs

△ *DOLERUS TRIPLICATUS* is a European species. Its larvae feed on the foliage of rushes that belong to the genus *Juncus*.

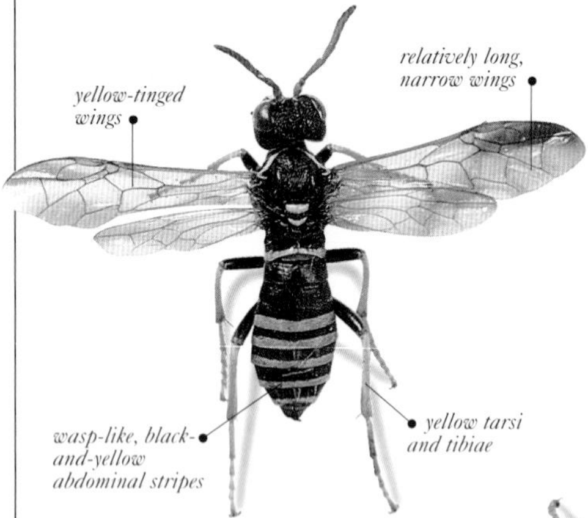

relatively long, narrow wings •

yellow-tinged wings •

◁ *TENTHREDO SCROPHULARIAE* is a wasp-mimicking species, found in Europe and Asia. Its larvae feed on mullein and figwort plants.

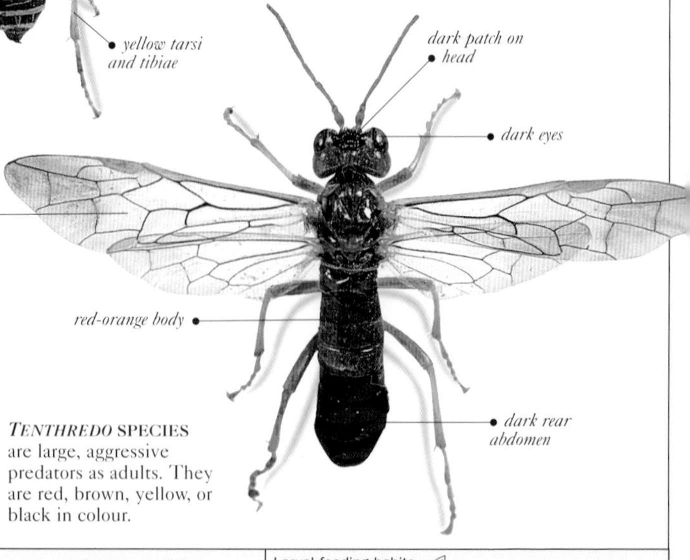

wasp-like, black-and-yellow abdominal stripes •

• yellow tarsi and tibiae

dark patch on • head

• dark eyes

yellow-tinged • wings

red-orange body •

• dark rear abdomen

LARVAE are mostly caterpillar-like, with a round head and abdominal prolegs.

TENTHREDO SPECIES are large, aggressive predators as adults. They are red, brown, yellow, or black in colour.

Length 0.3–2.2cm (⅛–⅞in), most under 1.6cm (⅝in)	Larval feeding habits

NON-INSECT HEXAPODS

SPRINGTAILS

C OMMONLY KNOWN as springtails, the order Collembola contains 18 families and 6,500 species. These small hexapods have a structure called a ventral tube on the underside of the abdomen. This is important in maintaining a salt and water balance and, in some species, for gripping smooth surfaces. Another feature is the jumping organ (furcula), which can be folded under the abdomen, where it engages with a catch. Muscular action releasing the furcula can throw the springtail well out of the way of predators.

Males deposit sperm on the ground or place it into the female's genital opening. Adulthood is reached after 5 to 13 moults, but adults continue to moult until they die. Springtails are vital in leaf-litter and soil chains, where hundreds of thousands may be found in one square metre.

Order COLLEMBOLA	Family ENTOMOBRYIDAE	No. of species 1,400

ENTOMOBRYIDS

These hexapods are pale to yellow, brown, or black in coloration. Some are patterned or mottled. They are elongate with a small pronotum, and in many the fourth abdominal segment is larger than the third segment. The antennae may be more than twice the body length.
• LIFE-CYCLE Females lay their eggs either in soil or in leaf-litter. All stages eat fungal threads or decaying plant matter.
• OCCURRENCE Worldwide. In leaf-litter, soil, and fungi in a variety of habitats. Some are found in caves.

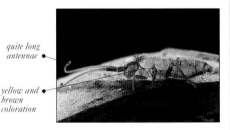

quite long antennae

yellow and brown coloration

ENTOMOBRYA SPECIES are commonly found on tree bark and rocks. Some species in this genus spend the whole of the winter on tree bark.

Length 1–8mm (½–⁵⁄₁₆in), most under 5mm (³⁄₁₆in)	Feeding habits

Order COLLEMBOLA	Family ISOTOMIDAE	No. of species 1,000

ISOTOMIDS

These springtails may be white, yellow, green, or brown in colour, and the upper surface is usually darker than the underside. The segments of the abdomen are equal in size.
• LIFE-CYCLE As in all springtails, males deposit rounded spermatophores on the ground, which the females take into their genital opening. A few isotomids are parthenogenetic.
• OCCURRENCE Worldwide. In soil in various habitats, but also around ponds and streams. A few species are abundant in harsh environments such as deserts, polar regions, and mountains.

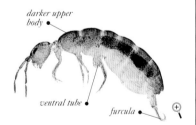

darker upper body

ventral tube

furcula

ISOTOMA VIRIDIS is often abundant among damp leaf-litter and moss clumps. The ventral tube and curved jumping organ are clearly visible on this specimen.

Length 1–8mm (½–⁵⁄₁₆in), most under 5mm (³⁄₁₆in)	Feeding habits

| Order COLLEMBOLA | Family NEANURIDAE | No. of species 1,000 |

NEANURID SPRINGTAILS

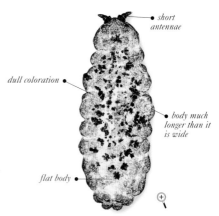

short antennae

dull coloration •

• body much longer than it is wide

flat body •

The body of most species in this family is longer than it is wide, but some may be squat or even flat. Many species are blue, grey, or red in colour, and a few have bands of contrasting colours. The body surface may be smooth or have blunt hairs or brightly coloured, hair-like projections.
• **LIFE-CYCLE** The eggs are laid in or under soil, leaf-litter, dung, stones, rotting wood, and bark. The young look much like small adults, and moulting continues after they have reached sexual maturity.
• **OCCURRENCE** Worldwide. In a variety of habitats, under stones and bark, in leaf-litter, soil, dung, and decaying wood.
• **REMARK** The dark blue species *Anurida maritima* is extremely common on seashores in the northern hemisphere, where it eats the remains of dead arthropods and snails. It survives by hiding inside air pockets that form between rocks during high tide.

NEANURA MUSCORUM is found worldwide, especially in woodland, under rotting wood and in soil. It is also found in caves. This species is able to produce chemicals to deter predatory spiders.

| Length 0.2–1cm (¹⁄₁₆–³⁄₈in) | Feeding habits |

| Order COLLEMBOLA | Family ONYCHIURIDAE | No. of species 600 |

BLIND SPRINGTAILS

Most members of this family are slender and pale or white. A few species have either blue-grey or slightly red coloration. As their common name suggests, the vast majority of blind springtails have no eyes. They do not have a furcula, although some species may have the vestigal remains of one. The body has a small number of thin-walled spots, or pores, on the cuticle of most segments, through which a noxious liquid can be secreted in order to deter predators.
• **LIFE-CYCLE** Eggs are laid in soil, leaf-litter, decaying wood, and fungi. The nymphs look like small adults, and moulting continues after they become sexually mature.
• **OCCURRENCE** Worldwide. In forests and pastures, and in caves, alpine areas, and even the Arctic. In soil, leaf-litter, rotting wood, and the fruiting bodies of fungi.

fine hairs on surface of body •

short, blunt antennae •

• *pale coloration*

ONYCHIURUS SPECIES are typical of the soil-dwelling springtails. Some species in this genus may be found in seashore habitats.

| Length 2–9mm (¹⁄₁₆–¹¹⁄₃₂in), most under 4mm (³⁄₃₂) | Feeding habits |

Order COLLEMBOLA	Family PODURIDAE	No. of species 1

THE WATER SPRINGTAIL

The single species in this family – *Podura aquatica* – is a very minute and common springtail. It varies in colour from brown or red-brown to dark blue or black. Its furcula is extremely well adapted for life on the water. It is quite flat and long, reaching the abdominal ventral tube (which helps the springtail to grip the water surface).

• **LIFE-CYCLE** This species spends much of its life scavenging on the surface of water. Its eggs are laid among vegetation found in and around bodies of water.

• **OCCURRENCE** Northern hemisphere. On the surface of fresh water in ditches, ponds, canals, and boggy areas.

• **REMARK** The furcula is particularly long in this species because a large area of it must be in contact with the elastic film that exists on the surface of water for the Water Springtail to jump effectively.

legs paler than body • *many individuals crowded together in sheltered area of pond* •

PODURA AQUATICA is well adapted to life on water. It is even found in puddles, especially in summer, and may gather in such large numbers that the puddle appears dark in colour, as if covered in soot.

Length Up to 2mm (¹⁄₁₆in)	Feeding habits

Order COLLEMBOLA	Family SMINTHURIDAE	No. of species 900

GLOBULAR SPRINGTAILS

Also known as garden springtails, these species are pale to dark brown or green in coloration, with spherical bodies. The segmentation on the abdomen is indistinct, and the antennae are noticeably long and elbowed. The males are often different in appearance to the females.

• **LIFE-CYCLE** In many males, the antennae are designed to hold the female during mating. Eggs are laid in small batches in soil, and development to sexual maturity may take as little as one month. There is some evidence of maternal care.

• **OCCURRENCE** Worldwide. In a wide variety of habitats, on trees, in leaf-litter, on the fruiting bodies of fungi, and on the surface of fresh water in ditches, bogs, and ponds. Also in damp places such as caves.

• **REMARK** Several species are significant pests of crop seedlings. *Sminthurus viridis*, which is commonly known as the Lucerne Flea, is a widespread pest of alfalfa and some vegetables – as many as 70,000 globular springtails have been recorded in just one square metre of pasture.

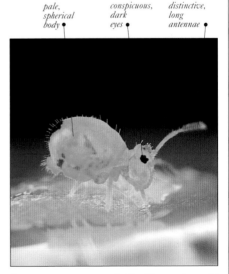

pale, spherical body • *conspicuous, dark eyes* • *distinctive, long antennae* •

SMINTHURIDES AQUATICUS is widespread on the surface of ponds and stagnant water but does not gather in such large numbers as the Water Springtail (see above).

Length 1–3mm (¹⁄₂₅–¹⁄₈in)	Feeding habits

PROTURANS

MEMBERS OF THE ORDER Protura, which contains 4 families and 400 species, are soil-dwelling hexapods. The first specimens were discovered in 1907.

These tiny creatures have neither eyes nor antennae, although there is a pair of tiny patches on either side of the head that may be the vestiges of antennae. In place of antennae, the front pair of legs are used as sensory organs. The middle and hind pairs are used for walking. Like springtails (see pp.207–209) and diplurans (see p.211), proturans have piercing-sucking mouthparts that are contained inside a pouch and pushed out whenever the animal is feeding. The jaws are sharp and rod-like in appearance. The first three abdominal segments may have minute vestiges of leg-like structures. There are no cerci.

During mating, sperm is transferred indirectly, with the male depositing a spermatophore on the ground that is picked up by the female's genitalia. When the larvae hatch out from the eggs, the abdomen has eight segments and a tail segment (telson). By the time they have moulted three times, proturans have the full complement of eleven abdominal segments plus the telson. Another two moults are required before they become sexually mature.

Order PROTURA	Family EOSENTOMIDAE	No. of species 90

EOSENTOMIDS

Pale and soft-bodied, these hexapods have a conical head and elongate body. The legs do not project far from the body, which allows easy passage through tiny cracks and crevices. The front legs are stouter than the middle and hindlegs and have numerous hairs and other sensory organs. Spiracles are visible on the middle and hind segments of the thorax.
• **LIFE-CYCLE** The eggs are round and patterned or have raised warts, and are usually laid in soil or leaf-litter. The larvae look much like small adults.
• **OCCURRENCE** Worldwide. In a variety of habitats, preferably in damp, cool conditions. Eosentomids occur in great numbers in soil, leaf-litter, moss, humus, and decaying wood.

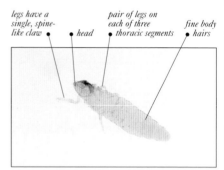

legs have a single, spine-like claw • *head* • *pair of legs on each of three thoracic segments* • *fine body hairs*

EOSENTOMON DELICATUM is native to Europe, and the genus as a whole is found all over the world. This species lives in soil, especially chalky soils.

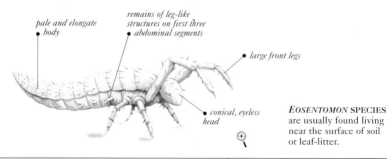

pale and elongate body • *remains of leg-like structures on first three abdominal segments* • *large front legs* • *conical, eyeless head*

EOSENTOMON SPECIES are usually found living near the surface of soil or leaf-litter.

Length 0.5–2mm (¹⁄₆₄–¹⁄₁₆in)	Feeding habits

DIPLURANS

THERE ARE 9 FAMILIES AND 800 species in the order Diplura. Pale in colour, these elongate, soft-bodied hexapods do not have eyes. They are sometimes called two-tailed bristletails, a name that refers to the two abdominal cerci, which may be long or pincer-like. They should not, however, be confused with the true bristletails (see p.46). The large head has long antennae and biting mouthparts contained within a pouch. Males deposit stalked spermatophores, which females take into their genital opening. Eggs are often laid in clumps, and females may guard their brood.

Diplurans live in rotting vegetation, compost heaps, and soil, and under stones and wood. With their slender, flexible bodies and strong legs, they can move through soil very easily.

Order DIPLURA	Family CAMPODEIDAE	No. of species 200

CAMPODEIDS

These white or yellow-tinged diplurans have long, multi-segmented cerci and supporting projections on the underside of their abdomen. Air is taken in through spiracles on the thorax.
• **LIFE-CYCLE** Eggs are usually laid in soil. Initially immobile, larvae become progressively more active and look like small adults.
• **OCCURRENCE** Worldwide. Widespread in various habitats, including caves. They are very common deep in soil, but are also found under tree bark and in decaying wood and vegetation.

CAMPODEA FRAGILIS is a common European and Asian species, found in rotting vegetation.

long, distinctive cerci

pale coloration

Length 0.4–1.2cm (⁵⁄₃₂–¹⁄₂in)	Feeding habits

Order DIPLURA	Family JAPYGIDAE	No. of species 200

JAPYGIDS

These species are pale, slender, and flexible, with telescopic antennae that can be shortened as they make their way through soil. The cerci are dark, tough, and forcep-like, similar to those of earwigs (see pp.69–70). Air is taken in through spiracles on the thorax and abdomen.
• **LIFE-CYCLE** Eggs are usually laid in soil. The young become more like the adults at successive moults. The abdominal cerci are used to catch small arthropod prey.
• **OCCURRENCE** Worldwide. In various habitats, in crevices in soil.
• **REMARK** Japygids can be distinguished from young earwigs by their lack of eyes.

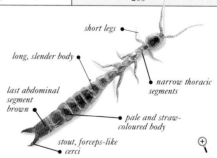

short legs

long, slender body

narrow thoracic segments

last abdominal segment brown

pale and straw-coloured body

stout, forceps-like cerci

HOLJAPYX DIVERSIUNGUIS, or the Slender Dipluran, is native to North America. It is a common soil-dwelling species, approximately 0.8–1cm (⁵⁄₁₆–³⁄₈in) in length.

Length 0.6–3cm (¹⁄₄–1¹⁄₄in)	Feeding habits

CRUSTACEANS

ISOPODS

T HE ORDER ISOPODA consists of 100 families and 10,000 species of crustacean. Most isopods are marine. However, 32 families (3,800 species) belong to a sub-order called Oniscoidea and are amphibious or live in terrestrial habitats. They are known collectively as woodlice. Woodlice have a segmented, flat body with seven pairs of similarly shaped and sized legs. The female carries the eggs inside a brood pouch, which is located beneath the abdomen. The young are kept in this pouch for a while after they hatch out.

Most woodlice favour damp and cool conditions, although some have become adapted to a wide range of habitats, including extremely dry regions. Some woodlice have camouflage colouring that blends with their background.

Order ISOPODA	Family ARMADILLIDIIDAE	No. of species 250

PILL WOODLICE

Also called pill bugs, these woodlice are convex in cross-section, with a rounded hind margin. Many can roll up into a ball for protection.
• **LIFE-CYCLE** The eggs are carried in a brood pouch until they hatch. As in many woodlice, the newly hatched young have one less pair of legs than the adults. The mother may provide her young with some protection.
• **OCCURRENCE** Europe and the Mediterranean. In places such as leaf-litter and debris.

• *curled up for protection*

• *segments of outer casing overlap*

ARMADILLIDIUM ALBUM is pale with dark markings and lives in salt marshes and coastal areas. It does not form as tight a ball as the pill millipede (see p.242) when it rolls up.

Length 0.5–2.5cm (⅕₆–1in)	Feeding habits

Order ISOPODA	Family PORCELLIONIDAE	No. of species 500

PORCELLIONIDS

The smooth or warty body surface of these woodlice is usually either grey or greyish brown, with various other markings. Some species of porcellionid have narrower bodies and are able to run quickly.
• **LIFE-CYCLE** The females carry the eggs in their brood pouch until they hatch.
• **OCCURRENCE** Worldwide, but mainly in temperate regions. Found mostly in leaf-litter and debris.
• **REMARK** As with all woodlice, nitrogenous wastes are excreted as ammonia gas – not as urine. This gives large colonies of woodlice a characteristic smell.

greyish brown body

warty body surface

PORCELLIO SCABER is a widespread species. It is usually a greyish colour, but orange and cream forms are often found in coastal regions.

Length 0.9–2cm (¹¹⁄₃₂–¾in)	Feeding habits

ARACHNIDS

SCORPIONS

T HE 9 FAMILIES AND 1,400 species of the order Scorpiones make up the most ancient group of all arachnids.

The cephalothorax carries four pairs of walking legs and large pedipalps with a pincer-like claw. There is a main pair of eyes situated centrally on the head and a variable number of pairs on the sides. The last segment of the mobile "tail" (telson) bears the sting and its poison gland, used to paralyze prey and for defence. The sting of some scorpions can be fatal to humans. Reproduction starts with complex courtship. After this, males deposit sperm on the ground that is picked up by the females' genitalia. Females bear live young that are carried on the mother's back until their first moult. Scorpions favour warm areas and hunt at night, hiding under stones by day.

Order SCORPIONES	Family BOTHRIURIDAE	No. of species 90

BOTHRIURIDS

These scorpions have a narrow cephalothorax. Unlike the members of some other scorpion families, their tibiae do not have spurs.
• LIFE-CYCLE Courtship can be lengthy and may involve stinging. As in all scorpions, eggs hatch out inside the female's body. The young are born live and climb on to their mother's back.
• OCCURRENCE South America, Australia, Africa, and the Himalayas. In dry and humid areas, often in burrows dug under stones and boulders.

CENTROMACHETES POCOCKI is one of three very similar Chilean species. It preys on field crickets and will also eat caterpillars.

short, stout claws

slender cephalothorax

dark brown body

Length 2.5–12cm (1–4¾in)	Feeding habits 🦗

Order SCORPIONES	Family BUTHIDAE	No. of species 520

BUTHIDS

In these scorpions, the sternum of the cephalothorax is roughly triangular. Some legs have spines on the tibiae.
• LIFE-CYCLE As in all scorpions, young are born live and climb on to the mother's back. Nymphs may take several years to mature.
• OCCURRENCE Worldwide. In a wide variety of habitats, ranging from desert to moist forest, where they are found in rock cracks and underneath stones, logs, and bark.
• REMARK Most of the dangerous scorpions belong to this family. Their powerful venom paralyzes the muscles, including those of the respiratory system and the heart.

rows of small, dark bumps along tail segments

BUTHUS OCCITANUS, common across the Mediterranean, is found on the ground, under rocks and logs.

pale claws with dark teeth

tibial spine

Length 0.8–12cm (⅜–4¾in)	Feeding habits 🦗

Order SCORPIONES	Family CHACTIDAE	No. of species 140

CHACTIDS

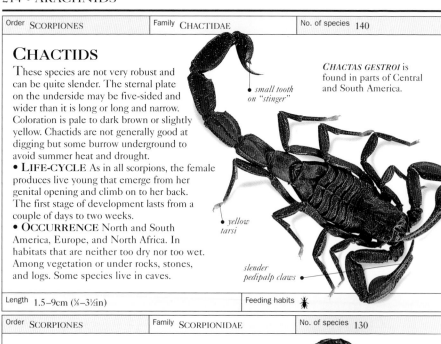

These species are not very robust and can be quite slender. The sternal plate on the underside may be five-sided and wider than it is long or long and narrow. Coloration is pale to dark brown or slightly yellow. Chactids are not generally good at digging but some burrow underground to avoid summer heat and drought.
• **LIFE-CYCLE** As in all scorpions, the female produces live young that emerge from her genital opening and climb on to her back. The first stage of development lasts from a couple of days to two weeks.
• **OCCURRENCE** North and South America, Europe, and North Africa. In habitats that are neither too dry nor too wet. Among vegetation or under rocks, stones, and logs. Some species live in caves.

CHACTAS GESTROI is found in parts of Central and South America.

small tooth on "stinger"

yellow tarsi

slender pedipalp claws

Length 1.5–9cm (⅝–3½in)		Feeding habits 🐜

Order SCORPIONES	Family SCORPIONIDAE	No. of species 130

SCORPIONIDS

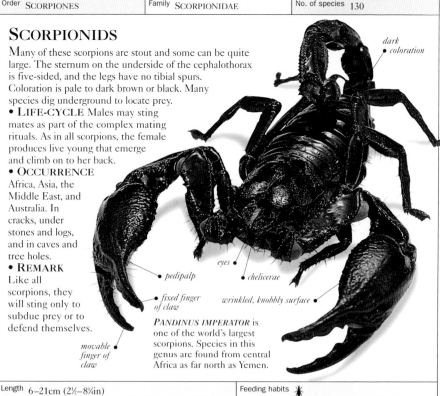

Many of these scorpions are stout and some can be quite large. The sternum on the underside of the cephalothorax is five-sided, and the legs have no tibial spurs. Coloration is pale to dark brown or black. Many species dig underground to locate prey.
• **LIFE-CYCLE** Males may sting mates as part of the complex mating rituals. As in all scorpions, the female produces live young that emerge and climb on to her back.
• **OCCURRENCE** Africa, Asia, the Middle East, and Australia. In cracks, under stones and logs, and in caves and tree holes.
• **REMARK** Like all scorpions, they will sting only to subdue prey or to defend themselves.

dark coloration

eyes

pedipalp

chelicerae

fixed finger of claw

wrinkled, knobbly surface

movable finger of claw

PANDINUS IMPERATOR is one of the world's largest scorpions. Species in this genus are found from central Africa as far north as Yemen.

Length 6–21cm (2½–8¼in)		Feeding habits 🐜

PSEUDOSCORPIONS

THE ORDER Pseudoscorpiones is divided into 23 families and 3,300 species. Also called false scorpions, they are similar in general shape to true scorpions, but are very small and lack the abdominal tail and sting of their larger relatives. The cephalothorax has a dorsal carapace, and the abdomen has 11 or 12 segments. Large, pincer-like pedipalps, which may be toothed, are used to catch prey and for defence, and the swollen parts of the pincers contain poison glands.

Males deposit sperm packets on the ground that are picked up by the females' genitalia. Eggs are laid into a pouch under the female's body. Pseudoscorpions make silk nests in which they moult, brood young, and hibernate. Most prefer moist or humid habitats – among leaf-litter or under stones, for example.

Order PSEUDOSCORPIONES	Family CHELIFERIDAE	No. of species 300

CHELIFERIDS

These pseudoscorpions have venom glands in both fingers of the pincers, and there are no teeth on the inner surfaces. They usually have two eyes. Coloration varies from pale to dark brown and black, in some cases tinged with red or olive, and with dark markings.
• **LIFE-CYCLE** Mating can be complex and may involve males and females dancing together, holding each other's pedipalps. As in all species, eggs are kept in a sac beneath the female and there are three nymphal stages. Some first-stage nymphs stay with the mother.
• **OCCURRENCE** Worldwide, especially in warmer regions. In leaf-litter and on tree bark.
• **REMARK** The species *Chelifer cancroides* is often found inside buildings.

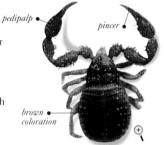

pedipalp • • pincer

brown coloration

DACTYLOCHELIFER SPECIES are found in parts of the northern hemisphere. Some species are confined to coastal habitats.

Length 1.5–5mm (¹⁄₁₆–³⁄₁₆in)	Feeding habits

Order PSEUDOSCORPIONES	Family CHERNETIDAE	No. of species 600

CHERNETIDS

In this family, the fingers of the pincers have teeth, and a poison gland is present only in the movable finger. The eyes are either weakly developed or absent. Males may be different in appearance from the females. Chernetids are shiny and coloured a variety of shades of brown.
• **LIFE-CYCLE** Males and females engage in a courtship dance, gripping each other with their pedipalps. As in all other species, the eggs are kept inside a sac under the female. Newly hatched nymphs may cling to the sides of their mother. There are three nymphal stages.
• **OCCURRENCE** Worldwide. In leaf-litter, debris, caves, and the nests and burrows of birds and small mammals.

pedipalp • • cephalothorax • abdomen

CHERNETID SPECIES are found in the northern hemisphere and in tropical regions of South America. The specimen shown here lives in caves in parts of Venezuela.

Length 1.5–5mm (¹⁄₁₆–³⁄₁₆in)	Feeding habits

Order PSEUDOSCORPIONES	Family CHTHONIIDAE	No. of species 570

CHTHONIIDS

In these arachnids, the abdomen is typically much longer than the carapace, which in turn covers the dorsal surface of the rest of the body and may be broader at the front than at the rear. Most species have four eyes and large chelicerae. On the first two pairs of legs, the tarsi have a single segment, whereas those on the third and fourth pairs of legs have two segments. The overall coloration varies from shades of brown to olive-green, and the legs are tinged with pink.

• LIFE-CYCLE Eggs are brooded inside the female's sac. The live young are usually released on to soil, leaf-litter, or bark. There are three nymphal stages, as in all pseudoscorpions.

• OCCURRENCE Worldwide, except in the extreme north or south. In various sheltered places, among soil and leaf-litter, and under tree bark. Others are found among seashore debris, in or near buildings, on wasteland, in caves, and in gardens and greenhouses.

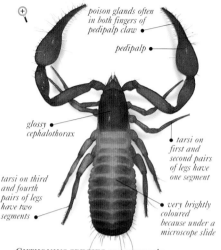

poison glands often in both fingers of pedipalp claw •

pedipalp •

glossy • cephalothorax

tarsi on first and second pairs of legs have one segment

tarsi on third and fourth pairs of legs have two segments •

• very brightly coloured because under a microscope slide

CHTHONIUS SPECIES are extremely widespread and are found in leaf-litter, at the base of grasses, under stones, and in the nests of birds and small mammals.

Length 1–2mm ($^1/_{32}$–$^1/_{16}$in)	Feeding habits 🐜

Order PSEUDOSCORPIONES	Family NEOBISIIDAE	No. of species 500

NEOBISIIDS

The carapace of these pseudoscorpions is quite angular or square when seen from above, and the chelicerae are large. In all the pairs of walking legs, the tarsi are made up of two segments, and the claw of the pedipalp has a poison gland only in the fixed finger. There are usually four eyes, but there may be fewer, or none at all, in cave-living species. The overall colouring varies from olive shades to dark brown, perhaps with red, yellow, or cream tinges. The legs are often slightly green in colour. Small items of prey are held and paralyzed with venom and are then shredded by the large chelicerae.

• LIFE-CYCLE Eggs are carried by the female in her brood sac, and the live young are typically released on to soil, leaf-litter, or bark. There are three nymphal stages.

• OCCURRENCE Worldwide, especially in the northern hemisphere. Many – typically smaller species – live in leaf-litter and soil, and some are found in caves.

red-brown pedipalps •   dark cephalothorax •   • olive-green legs

NEOBISIUM MARITIMUM is native to coastal areas of Ireland, England, and France. It is found in cracks in rocks and under stones, from the upper shore to the splash zone.

Length 1–5mm ($^1/_{32}$–$^3/_{16}$in)	Feeding habits 🐜

SUN-SPIDERS

D ESPITE RESEMBLING SPIDERS and scorpions, the 12 families and 1,000 species of Solifugae form a separate order. Also called wind-scorpions, sun-spiders have a flexible abdomen that narrows where it joins the three-sectioned cephalothorax and a head that bears a pair of small eyes. All sun-spiders are predatory. They use huge, pincer-like chelicerae to kill and macerate their prey.

The clawless, leg-like pedipalps have suction pads that enable them to grasp small vertebrates and arthropods. Sun-spiders are equipped with many sensitive body hairs and organs at the bases of the last pair of walking legs.

Most species are found in Southeast Asia, Africa, and North America. Males push sperm into the female's genital opening, and eggs are laid in a burrow.

Order SOLIFUGAE	Family AMMOTRECHIDAE	No. of species 72

AMMOTRECHIDS

These relatively slender species are quite variable and may be coloured in a range of shades of brown. The tarsi of the first pair of legs have no claws, and the front margin of the head is rounded. Many species are nocturnal, but some are active during the day. Nocturnal species typically dig into soil during the day, but some of the smaller species hide inside termite colonies or the tunnels of wood-boring insects.
• LIFE-CYCLE The female lays her eggs, perhaps in several batches, in a burrow.
• OCCURRENCE Warmer regions of Central America, South America (as far as southern Argentina), and North America. In dry areas, semi-deserts, and deserts.

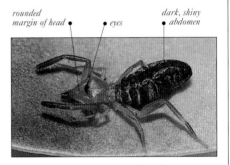

rounded margin of head • • *eyes* *dark, shiny* • *abdomen*

AMMOTRECHELLA STIMPSONI is native to Florida. This species is found underneath the bark of rotten trees and feeds on a variety of prey, including termites.

Length 0.4–2cm (⁵⁄₃₂–³⁄₄in)	Feeding habits 🦗

Order SOLIFUGAE	Family EREMOBATIDAE	No. of species 120

EREMOBATIDS

These species can be robust with short legs, or slender with long legs. The tarsi of the first three pairs of legs have one segment, whereas those of the fourth pair may have one to three segments. The front of the head looks square-cut. Coloration varies between light and dark brown.
• LIFE-CYCLE Eggs are laid in burrows. The young are especially fond of termite prey.
• OCCURRENCE Warm, dry parts of Central America and southern North America. In dry areas, semi-deserts and deserts, and mountainous regions.

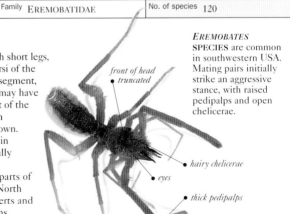

EREMOBATES SPECIES are common in southwestern USA. Mating pairs initially strike an aggressive stance, with raised pedipalps and open chelicerae.

front of head • *truncated*

hairy chelicerae

eyes

thick pedipalps

Length 0.8–4cm (⁵⁄₁₆–1¹⁄₂in)	Feeding habits 🦗

Order SOLIFUGAE	Family GALEODIDAE	No. of species 180

GALEODIDS

Members of this family have yellow, pale brown, red-tinged, or dark bodies. The tarsi on the first pair of legs have one segment, those on the second and third pairs have two segments, and those on the fourth pair have three segments. The claws on the last three pairs of legs are hairy. Galeodids hunt after dark, hiding away from the heat of the day in burrows that they have dug into sandy soil.
• **LIFE-CYCLE** Mating may involve the male carrying the female. Like all sun spiders, males place a spermatophore into the female's genital opening. Eggs are laid in a pit or burrow.
• **OCCURRENCE** Asia and northern Africa. In semi-arid and desert regions.
• **REMARK** Large species kill and eat lizards.

leg-like pedipalp •   • soft, flat abdomen

GALEODES CITRINUS frequently has yellow overall coloration. Here, the female is seen making a shallow pit in the ground, in which she will then lay her eggs.

• velvety abdomen

• eyes

• chelicerae

• pedipalp has no claw at end

GALEODES ARABS is a common North African species. It typically jumps on passing prey and then retires into its burrow to rest and digest its food.

Length 1–7.2cm (⅜–2¾in)	Feeding habits 🦗

Order SOLIFUGAE	Family SOLPUGIDAE	No. of species 200

SOLPUGIDS

These sun spiders are pale straw-coloured, brown, or slightly yellow, and some have bright markings. The tarsi on the first pair of legs have one segment, those on the second and third pairs have four segments, and those on the last pair have six or seven segments. All but the first pair of legs have smooth-surfaced claws. Some species are active by day; others hide in burrows dug in sandy soil, in cracks, or under stones.
• **LIFE-CYCLE** The females lay their eggs inside pits in the ground. Small solpugids and nymphs feed on termites.
• **OCCURRENCE** Africa and parts of the Middle East. In woodland, dry savannah, and semi-arid and desert regions.

large orange-red   central brown mark   extensive covering
cephalothorax •   on abdomen •   • of white hairs

METASOLPUGA PICTA is found in the Namib desert in southern Africa. Like all solpugids, it will bite if handled carelessly, but does not have poison glands.

Length 0.6–6cm (¼–2½in)	Feeding habits 🦗

WHIP-SCORPIONS

T HE ORDER UROPYGI consists of just 2 families comprising 99 species. These flattened arachnids are also known as vinegaroons because of their ability to defend themselves by spraying formic and acetic acids from a pair of glands at the end of their abdomen. The cephalothorax is longer than it is wide. It is covered by a carapace that carries a pair of eyes at the front edge and several eyes on each side. The chelicerae are more like spider fangs than pincers, and the abdomen has 12 segments and ends in a whip-like "tail" section that is quite different from that of true scorpions. The robust pedipalps are used to catch, hold, and crush prey. Reproduction is similar to that of true scorpions, and females carry hatched young on their backs.

Order UROPYGI	Family THELYPHONIDAE	No. of species 75

VINEGAROONS

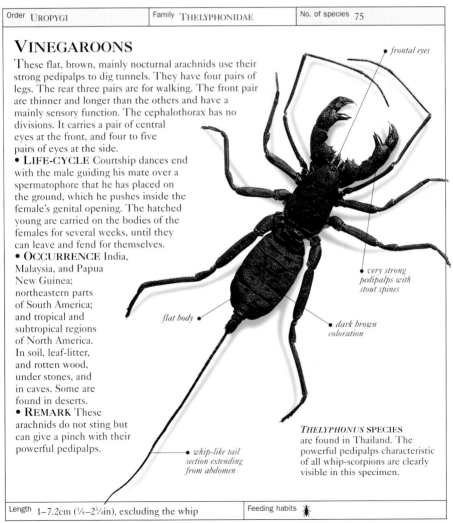

These flat, brown, mainly nocturnal arachnids use their strong pedipalps to dig tunnels. They have four pairs of legs. The rear three pairs are for walking. The front pair are thinner and longer than the others and have a mainly sensory function. The cephalothorax has no divisions. It carries a pair of central eyes at the front, and four to five pairs of eyes at the side.

• **LIFE-CYCLE** Courtship dances end with the male guiding his mate over a spermatophore that he has placed on the ground, which he pushes inside the female's genital opening. The hatched young are carried on the bodies of the females for several weeks, until they can leave and fend for themselves.

• **OCCURRENCE** India, Malaysia, and Papua New Guinea; northeastern parts of South America; and tropical and subtropical regions of North America. In soil, leaf-litter, and rotten wood, under stones, and in caves. Some are found in deserts.

• **REMARK** These arachnids do not sting but can give a pinch with their powerful pedipalps.

frontal eyes

very strong pedipalps with stout spines

flat body

dark brown coloration

whip-like tail section extending from abdomen

***THELYPHONUS* SPECIES** are found in Thailand. The powerful pedipalps characteristic of all whip-scorpions are clearly visible in this specimen.

Length 1–7.2cm (³⁄₈–2³⁄₄in), excluding the whip	Feeding habits 🐜

WHIP-SPIDERS

T HE ORDER AMBLYPYGI is divided into 3 families and 130 species. Also called tailless whip-scorpions, these arachnids have squat bodies, which are flat in profile, and a broad cephalothorax. The first segment of the rounded abdomen is stalk-like. Whip-spiders have eight eyes: a middle pair and three lateral pairs. The large pedipalps may be long and slender or short and stout. Spiny, sharp-tipped, and six-segmented, they seize and hold prey, while the two-segmented, fang-like chelicerae tear pieces off. The much-segmented, very long first pair of legs are used as feelers.

Whip-spiders are nocturnal, do not sting or bite, and prey on insects and other arthropods. Eggs hatch out inside a sac under the female's abdomen, and the young then climb on to her back.

Order AMBLYPYGI	Family PHRYNIDAE	No. of species 52

PHRYNIDS

These whip-spiders are generally coloured various shades of brown, with darker markings. The tibia of the fourth pair of legs is divided into three or four segments. In some species, the pedipalps of the males are proportionately longer than those of the females.
• LIFE-CYCLE As with all members of this order, males deposit a spermatophore on the ground, from where it is taken up by the female's genitalia. Females carry their eggs and then the hatched young inside a brood sac. The young later climb on to their mother's back and are carried around until they are able to look after themselves.
• OCCURRENCE Tropical and subtropical areas. In moist places: under bark and in leaf-litter in wooded areas; among stones and rocks; and in caves.

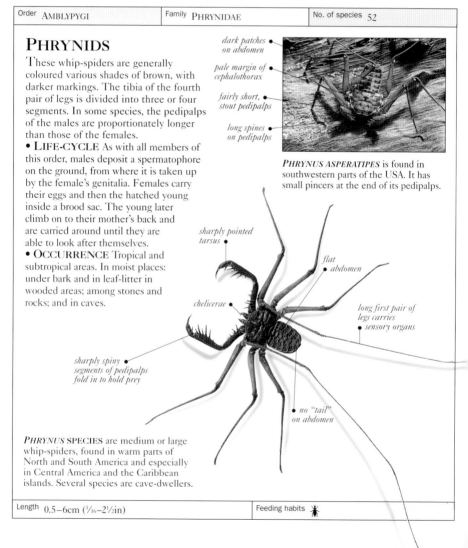

dark patches on abdomen

pale margin of cephalothorax

fairly short, stout pedipalps

long spines on pedipalps

PHRYNUS ASPERATIPES is found in southwestern parts of the USA. It has small pincers at the end of its pedipalps.

sharply pointed tarsus

flat abdomen

chelicerae

long first pair of legs carries sensory organs

sharply spiny segments of pedipalps fold in to hold prey

no "tail" on abdomen

PHRYNUS SPECIES are medium or large whip-spiders, found in warm parts of North and South America and especially in Central America and the Caribbean islands. Several species are cave-dwellers.

Length 0.5–6cm (³⁄₁₆–2¹⁄₂in)	Feeding habits 🦗

HARVESTMEN

OMMONLY known as harvestmen, the order Opiliones contains 40 families and 5,000 species. They lack a slender waist between their cephalothorax and their abdomen. They have a pair of eyes at the front of the cephalothorax, often carried on a raised structure. The pincer-like chelicerae have three segments, and the pedipalps have six. Their legs can be short or long.

The four pairs of walking legs have seven segments and usually one or two simple claws. Special glands in the cephalothorax produce smelly secretions, which are used as a defence.

Harvestmen are unusual among arachnids in that fertilization is direct – males have a penis for transferring sperm. Females may have an ovipositor with which they lay eggs in cracks in the soil.

Order OPILIONES	Family COSMETIDAE	No. of species 450

COSMETIDS

Like most harvestmen, these species are generally dull-coloured. Some tropical species, however, are green or yellow, and a few are able to change colour to blend in with their background.
• LIFE-CYCLE As for the whole order, fertilization is direct. Eggs are deposited in soil or other damp, sheltered spots.
• OCCURRENCE Mainly tropical regions of North and South America. Under stones and among debris in grassland, forest, and semi-desert.

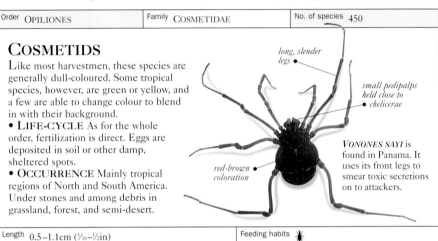

long, slender legs

small pedipalps held close to chelicerae

red-brown coloration

VONONES SAYI is found in Panama. It uses its front legs to smear toxic secretions on to attackers.

Length 0.5–1.1cm (³⁄₁₆–½in)	Feeding habits

Order OPILIONES	Family GONYLEPTIDAE	No. of species 750

GONYLEPTIDS

Members of this family typically have stout bodies with a broad, sometimes flat-ended, rear. Many are brightly coloured. The first part of the hindleg is enlarged and may have long, sharp spines. The eyes are close together and borne on a small protuberance. Males tend to have smaller bodies, and often much spinier legs, than females. Most species are active after dark and may produce chemicals to deter attackers.
• LIFE-CYCLE Eggs are laid in damp, sheltered spots. Generally, the females do not look after their eggs, although there is one species that builds a protective mud wall around both herself and her eggs.
• OCCURRENCE Mainly in South American tropical forests. Under logs and stones.

claw at tip of pedipalp

small bumps on surface of body

first segment of fourth leg enlarged

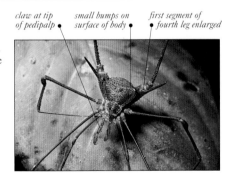

DISCOCYRTUS SPECIES are natives of the Brazilian rainforests. With their distinctive triangular bodies and enlarged hindlegs, their appearance is typical of this family.

Length 0.5–1.4cm (³⁄₁₆–⅝in)	Feeding habits

Order OPILIONES	Family LEIOBUNIDAE	No. of species 450

LEIOBUNIDS

The bodies of these harvestmen vary. Most have very long, slender legs, with two rows of small "teeth" on the first segment. The second pair of walking legs may be 15 times as long as the body.

• **LIFE-CYCLE** Little is known about courtship and egg-laying. In some species, mating involves large gatherings of males and females on tree stumps or mossy knolls, where males fight each other, often biting off each other's legs. Larger males usually win the contests and mate with the waiting females.

• **OCCURRENCE** Temperate parts of the northern hemisphere, especially North America and Europe, and in some tropical regions but absent from Africa. In moist places in woodland and cave entrances.

• **REMARK** Leiobunids use their eyes to distinguish light and dark, but they are not able to perceive images.

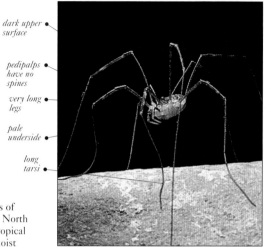

dark upper surface

pedipalps have no spines

very long legs

pale underside

long tarsi

LEIOBUNUM ROTUNDUM is active at night, descending from trees to hunt for food at ground level. Its long, flexible tarsi can be wrapped tightly around grass blades for a strong grip.

Length 0.2–1.2cm (¹⁄₁₆–¹⁄₂in), body only	Feeding habits

Order OPILIONES	Family PHALANGIIDAE	No. of species 200

PHALANGIIDS

These arachnids usually have soft bodies and may have many spiny projections. The first leg segment is smooth, but the other segments may have longitudinal, sometimes spined, ridges. Males and females may differ, the male's enlarged chelicerae being especially distinct. Many species are nocturnal, but some are also active during the day.

• **LIFE-CYCLE** Females use their telescopic, flexible ovipositor to lay eggs under bark or in soil crevices. The young stay in low vegetation at first, climbing into bushes and trees when older.

• **OCCURRENCE** Worldwide, mainly in temperate regions. Under stones and among leaf-litter in wooded and grassy areas.

• **REMARK** Several species are now adapted to living in houses.

tarsus of second leg may have 50 segments

second pair of legs is especially long

saddle-shaped mark

PHALANGIUM OPILIO is a white-grey to yellow species with a saddle-shaped mark on its back. It is found in the woods, gardens, and grasslands of the northern hemisphere.

eye

Length 0.2–1.2cm (¹⁄₁₆–¹⁄₂in), body only	Feeding habits

TICKS AND MITES

T ICKS AND MITES form the order Acari, a huge, diverse group of about 300 families and 30,000 species. They are found in every habitat, including aquatic ones, and have a wide range of lifestyles. Many are significant pests of crops and stored produce or parasitize humans and other animals.

Most species are less than 1mm (½in) long, although ticks can be much larger, especially following a blood meal. The body has no distinctive divisions, and the short abdomen has no segments. The mouthparts are carried on a special extension. The chelicerae are two- or three-segmented pincers or are adapted for piercing and sucking. Both the adults and nymphs have four pairs of six-segmented walking legs, although the first-stage larvae have only three pairs.

Order ACARI	Family ACARIDAE	No. of species 550

ACARIDS

Also called storage mites, most acarids are pale in coloration. The abdomen has long hairs. The legs of these mites can be long but in some species are extremely short.
• LIFE-CYCLE Eggs are laid wherever the mites feed – see Occurrence. As with most mites, there are three nymphal stages. Many live in association with certain arthropods, and some are found in rotting matter.
• OCCURRENCE Worldwide. In fresh or dried stored products, cheese, fungi, bee hives, organic detritus, and inside mattresses.
• REMARK A few acarids are pests of dried, stored food. Some eat mammalian skin or bite humans and can cause skin conditions such as dermatitis or trigger allergies such as asthma.

long body hairs • translucent body •

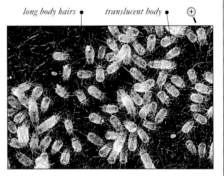

ACARUS SIRO, the Flour Mite, is found all over the world in flour, grain, and various seeds in stores and mills. If conditions are suitable, huge populations can build up.

Length Under 1mm (½in)	Feeding habits

Order ACARI	Family ARGASIDAE	No. of species 150

SOFT TICKS

These ticks usually have a rounded, berry-like body, although some are flat dorsoventrally. The tough, leathery body can be either wrinkled or folded, and the chelicerae are adapted for cutting through the skin of their hosts – mammals (including bats), birds, and snakes. They are ectoparasitic and feed mostly at night.
• LIFE-CYCLE Eggs are typically laid in the nests and burrows of their hosts, and both adults and nymphs live mainly in association with these animals.
• OCCURRENCE Worldwide, especially in warm, dry regions. Typically in hosts' nests and burrows.
• REMARK Many soft ticks are carriers of disease and are pests of various domestic animals, especially poultry.

• pedipalps

• pale, curved legs

leathery body •

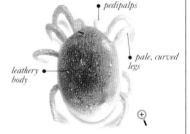

ARGAS PERSICUS is a pest of domestic chickens and other poultry in many parts of the world. It transmits a disease called fowl relapsing fever.

Length 0.2–1cm (¹⁄₁₆–³⁄₈in), most under 0.6cm (¼in)	Feeding habits

Order ACARI	Family DERMANYSSIDAE	No. of species 25

DERMANYSSIDS

These mites use their needle-like chelicerae to feed on the blood of birds and mammals. After a blood meal, their colour changes from pale grey to red. Many females have a single dorsal plate with short hairs.

translucent body turns red after a blood meal

pedipalps

four pairs of similarly sized legs

• **LIFE-CYCLE** Males use their chelicerae to transfer sperm to the female. Eggs are laid in places such as nests, burrows, and poultry houses. The first-stage larva does not feed, although subsequent nymphal stages do.
• **OCCURRENCE** Worldwide. In association with bird and mammal hosts.
• **REMARK** Certain species are significant pests of poultry, and some carry diseases that can kill animals and also affect humans.

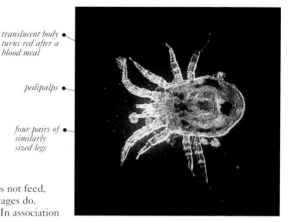

DERMANYSSUS GALLINAE, the Red Poultry Mite, is found all over the world on a wide range of birds. These mites feed at night and hide in crevices during the day.

Length 0.2–0.8mm (less than ¹⁄₃₂in)	Feeding habits

Order ACARI	Family IXODIDAE	No. of species 650

HARD TICKS

These flat ticks have a very tough, sometimes patterned plate on their back. In males, it covers the whole body; in females and immature ticks, it covers only the front half. The soft, flexible abdomen allows large blood meals to be taken from the animal hosts on which these ticks are found. Colouring varies from yellow to red- or black-brown, and some species are highly marked.

mouthparts

palp

front part of head ("false head") projects forwards

in females, dorsal plate covers only front half of body

• **LIFE-CYCLE** After mating, a female gorges herself on blood and then drops off the host to lay a batch of eggs among vegetation. Six-legged larvae emerge, crawl up grass blades, and attach themselves to a passing host. A larva feeds for a few days and then drops off the host to moult into an eight-legged nymph. The nymph attaches itself to a host and feeds for several days before once again dropping off to moult into an adult.
• **OCCURRENCE** Worldwide. In association with bird, mammal, and some reptile hosts.
• **REMARK** Many hard ticks transmit disease and are serious pests of domestic animals such as cattle, sheep, horses, and poultry. Some also carry viral diseases that affect humans, including encephalitis, Lyme disease, tick typhus, and Rocky Mountain Spotted Fever.

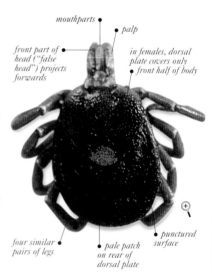

four similar pairs of legs

pale patch on rear of dorsal plate

punctured surface

AMBLYOMMA AMERICANUM, the Lone Star Tick, attacks a range of mammals and will also bite humans. It is found in the central states of the USA.

Length 0.2–1cm (¹⁄₁₆–³⁄₈in); larger when engorged	Feeding habits

Order ACARI	Family LAELAPIDAE	No. of species 650

LAELAPID MITES

These brown mites are ectoparasites of insects or mammals. The former have weak, hair-like structures on the body. In the latter, these are spinier, helping the mites to cling to their hosts. The dorsal plate is not divided in two, as it is in some mites.
• LIFE-CYCLE Males transfer sperm to females with their chelicerae. Many species feed on the lymph or blood of mammals and lay eggs in the host's nest or burrow. Some produce live larvae.
• OCCURRENCE Worldwide. In a wide variety of habitats: in poultry houses, the nests of small animals, and ant colonies, and in dung, tidal debris, and stored produce.
• REMARK Many species transmit disease.

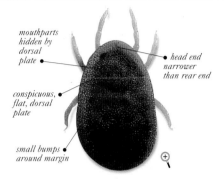

mouthparts hidden by dorsal plate

head end narrower than rear end

conspicuous, flat, dorsal plate

small bumps around margin

HAEMOLAELAPS GLASGOWI is widespread on rats. It can transmit the virus that causes epidemic haemorrhagic fever between rats and other rodents and possibly also humans.

Length 0.5–5mm (⅟₆₄–⅟₁₆in), most under 2mm (⅟₁₆in)	Feeding habits

Order ACARI	Family MICROTROMBIDIIDAE	No. of species 500

MICROTROMBIDIIDS

Microtrombidiids are usually brown and densely hairy. The legs have six segments and the front of the dorsal plate always carries two pairs of eyes.
• LIFE-CYCLE These mites parasitize other arthropods. The females lay up to 4,000 eggs in soil, and their hatched larvae feed off a suitable host. The larvae then moult into nymphs, eat insect eggs that they find in the soil, and develop into adults.
• OCCURRENCE Worldwide. In various habitats, especially dry, sandy, or semi-arid areas.

yellow-cream colour

UNDERSIDE

soft, velvety hair

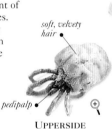

pedipalp

UPPERSIDE

EUTROMBIDIUM SPECIES, like the one shown here, are usually found on the body surface of praying mantids, crickets, grasshoppers, and locusts.

Length 0.5–2mm (⅟₆₄–⅟₁₆in)	Feeding habits

Order ACARI	Family PARASITIDAE	No. of species 375

PARASITID MITES

Most of these mites are slightly pear-shaped and yellow-brown, with one or two visible plates on the dorsal surface. In males, the second pair of legs may be stouter and adapted to grasp females when mating.
• LIFE-CYCLE Eggs are laid in organic debris. Nymphs are often found on insects, and many eat small insects, their larvae, and other mites.
• OCCURRENCE Worldwide. In dung, wood, and plant debris, on other mites in stored produce, and in mammal, bee, and wasp nests.

shiny, smooth surface

rounded body

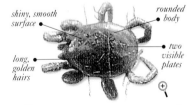

long, golden hairs

two visible plates

PARASITUS SPECIES are common and can be found inside the nests of wasps and bees, as well as among decaying wood and leaf-litter.

Length 0.75–2mm (⅟₆₄–⅟₁₆in)	Feeding habits

Order ACARI	Family SARCOPTIDAE	No. of species 120

SCABIES MITES

These small mites, also known as mange mites, are a pale, translucent brown. They have short, compact legs and almost spherical bodies that are slightly flat in profile. Their chelicerae are adapted for cutting the skin of their animal and human hosts (infesting humans with scabies and animals with mange).

• **LIFE-CYCLE** Most species feed on the host's epidermis and lymph, leaving tunnels in the skin. Mating occurs on the skin, and females lay up to 50 eggs in the tunnels during their lifetime. The hatched young find shelter and food in hair follicles.

• **OCCURRENCE** Worldwide. In the skin or hair follicles of mammals, including humans.

• **REMARK** Infestation causes extreme itching. Scratching leads to hair loss, and serious secondary infections can follow.

short, stout legs •   *long, fine body hairs* •   • *rounded outline*

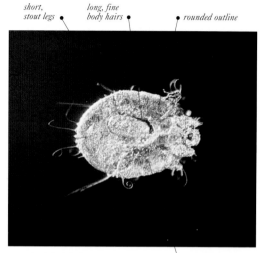

SARCOPTES SCABEI mites are the commonest cause of mange. There are many varieties within this species, each linked with a particular host.

• *fine transverse wrinkles on body*

Length 0.2–0.4mm (¹⁄₁₂₈–¹⁄₆₄in)	Feeding habits

Order ACARI	Family TETRANYCHIDAE	No. of species 650

SPIDER MITES

These mites are orange, red, green, or yellow in colour, with spider-like bodies. Large numbers feed on and infest host plants, which may then wither and develop pale blotches. Spider mites produce silk from glands at the front of their body and often cover affected plant parts with a fine webbing.

• **LIFE-CYCLE** Red, rounded, quite large eggs are laid on the leaves, twigs, or bark of host plants. The mites live under the leaves, protected from harm by their silk webs.

• **OCCURRENCE** Worldwide. On a range of plants, trees, and shrubs.

• **REMARK** Many spider mites are significant pests of grasses and other plants. Affected crops include wheat, citrus and other fruit trees, clover, cotton, and coffee. Infestation can seriously affect crop yields.

soft, rounded, or pear-shaped body •   *spider-like appearance* •   • *orange-red body*

TETRANYCHUS SPECIES feed on a wide range of plants and spend the winter deep in leaf-litter, emerging in the spring to locate host plants.

• *fine, pale body hairs*

Length 0.2–0.8mm (¹⁄₁₂₈–¹⁄₃₂in)	Feeding habits

Order ACARI	Family TROMBICULIDAE	No. of species 3,000

CHIGGER MITES

These mites are pale to mid brown or sometimes red. They are oval or slightly constricted in the middle, and the body and legs may have quite long hairs, although some have a velvety surface. Chigger mites parasitize mammals (including humans), reptiles, and birds.
• **LIFE-CYCLE** Eggs are laid in damp soil and larvae climb grass blades to find passing hosts. First-stage larvae feed on the outside of mammals, birds, snakes, and lizards, penetrating the skin with saw-like chelicerae to eat lymph and tissue. A few species feed in the tracheal system. When fully fed, the larva drops off, moults, and preys on small arthropods such as springtails (see pp.207–209).
• **OCCURRENCE** Worldwide. In soil, leaf-litter, and animals' burrows, or on hosts.
• **REMARK** Species that attack humans cause severe itching, dermatitis, and allergic reactions. A few carry scrub typhus from rodents to humans.

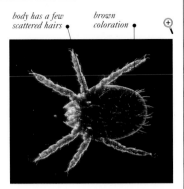

body has a few scattered hairs

brown coloration

NEOTROMBICULA AUTUMNALIS lives in soil and emerges on to the surface when it is warm and wet. Immature stages bite birds and mammals (including humans).

Length 1–3mm (¹⁄₃₂–⅛in)	Feeding habits

Order ACARI	Family TROMBIDIIDAE	No. of species 250

VELVET MITES

Many velvet mites have red or orange bodies that are extremely hairy, giving them a dense, velvety appearance. The body is not constricted in the middle.
• **LIFE-CYCLE** At certain times of year, often after rain, adults emerge from the soil to mate and lay eggs. Some larvae are parasites on insects, such as grasshoppers, and other arthropods.
• **OCCURRENCE** Worldwide, especially numerous in tropical regions. In various terrestrial habitats, from savannah to forests, mostly in or on soil. Some species are associated with fresh water.

dense, velvety covering of hairs

bumpy surface

TROMBIDIUM SPECIES are often seen walking over bare ground, especially after heavy rain, when they are forced from the soil in large numbers.

Length 0.2–1cm (¹⁄₁₆–⅜in), most under 0.5cm (³⁄₁₆in)	Feeding habits

Order ACARI	Family VARROIDAE	No. of species 5

VARROA MITES

Typically, varroa mites are pale tan in colour and broader than they are long, with smooth, oval, slightly convex bodies. They parasitize bees.
• **LIFE-CYCLE** Eggs are laid in bees' brood cells and the nymphs feed off the bee larvae. Adult mites attach themselves to adult bees, in order to feed off them and as a way of dispersing.
• **OCCURRENCE** Worldwide. Where hosts occur.

smooth, oval outline

VARROA PERSICUS attaches itself to the bodies of both wild and domestic honeybees.

legs

mouthparts

Length 1–1.75mm (¹⁄₃₂–¹⁄₁₆in)	Feeding habits

SPIDERS

MEMBERS OF THE 101 families and 40,000 species in the order Araneae are distinguished by their general appearance and their ability to spin silk thread and make webs.

The cephalothorax is covered by a carapace and is joined to the abdomen by a stalk. The front part of the carapace carries the eyes. Most species have eight simple eyes, but some have six, four, or two eyes, or none at all. The chelicerae have a hinged fang at the tip, and almost all species have poison glands. A spider's pedipalps are six-segmented and have a sensory function. In males, they are also used to transfer sperm. There are four pairs of seven-segmented walking legs. The abdomen is not segmented and carries silk-spinning organs (known as spinnerets) and a genital opening called the epigyne. When a spider feeds, the body tissues of its prey are dissolved by enzymes in the spider's digestive juices, producing a liquid that it then sucks up. Typically, the round spider eggs are laid inside a silk sac, which some species carry until the young hatch.

Spiders are found in almost every terrestrial habitat, from deserts to mountain peaks. They cannot fly, but many are able to travel long distances by "ballooning" on silk threads.

Order ARANEAE	Family AGELENIDAE	No. of species 700

FUNNEL WEAVERS

These spiders have hairy bodies and often have long legs. The narrow front of the cephalothorax bears eight eyes and the oval, quite slender abdomen may have dark bars, chevrons, or spots. The two posterior spinnerets have two segments and are longer than the anterior ones. Typically, these spiders make a funnel-shaped retreat at the margin of a flat web.

• **LIFE-CYCLE** After mating, males and females may stay together until the male dies. The egg sac is covered with silk and debris and is kept in the web. The young may be fed with regurgitated food.

• **OCCURRENCE** Worldwide. In various habitats, including grassland, meadows, and gardens. Webs are made in bushes, among stones, on rocks and walls, under logs, and inside houses.

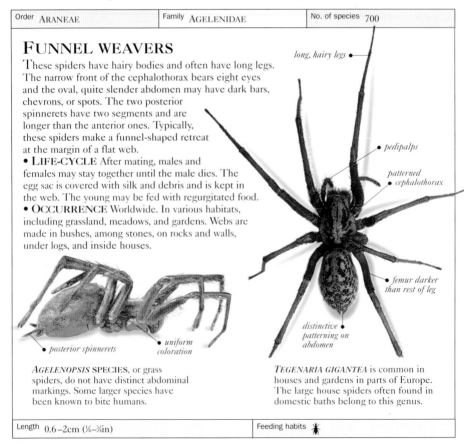

long, hairy legs

pedipalps

patterned cephalothorax

femur darker than rest of leg

distinctive patterning on abdomen

posterior spinnerets

uniform coloration

AGELENOPSIS SPECIES, or grass spiders, do not have distinct abdominal markings. Some larger species have been known to bite humans.

TEGENARIA GIGANTEA is common in houses and gardens in parts of Europe. The large house spiders often found in domestic baths belong to this genus.

Length 0.6–2cm (¼–¾in)		Feeding habits 🦗

Order ARANEAE	Family ARANEIDAE	No. of species 4,000

ORB WEB SPIDERS

These spiders often have very large abdomens, which can be brightly coloured and patterned. In some species, the abdomen may have a strange, angular shape. The legs have three claws and can be very spiny. They have eight eyes – the middle four often forming a square. Males are often smaller than females. The webs often have a central hub with radiating lines and spirals. Certain species do not make webs at all. Instead, they ensnare moths after dark using a single thread with a bead of glue at the end.
• **LIFE-CYCLE** Mating involves complex courtship. Silk egg sacs are kept camouflaged inside the web, stuck to vegetation or bark, or buried in leaf-litter.
• **OCCURRENCE** Worldwide. In a wide variety of different habitats, including grassland, meadows, forests, and gardens.
• **REMARK** Some tropical species make huge, strong webs and have been known to catch and eat birds. The enormous webs of *Nephila* species are used as fishing nets in Papua New Guinea.

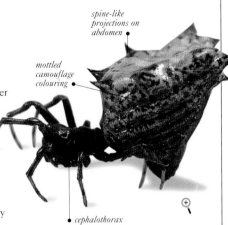

spine-like projections on abdomen

mottled camouflage colouring

cephalothorax

MICRATHENA GRACILIS, the spiny orb-weaving spider, is found in North American deciduous forests. The genus has odd, spiny protrusions on the abdomen.

Length 0.2–4.6cm (¹⁄₁₆–1¾in)	Feeding habits 🦗

Order ARANEAE	Family ARGYRONETIDAE	No. of species 1

THE WATER SPIDER

There is just one species in this family – *Argyroneta aquatica*. It is aptly named, as it lives more or less permanently under water. This spider has a distinctive, dense pile of short hairs on its grey abdomen. The legs are yellow-brown, and the third and fourth pairs have extensive tufts of longer hairs, which help to trap air. The Water Spider makes a dome-shaped "diving bell" out of a sheet of silk, which it attaches to submerged vegetation and fills with air. Bubbles of air are carried from the surface using the abdomen and hindlegs, and the air is 'brushed" off by the legs to fill the bell. The spider stays inside the bell with its long legs hanging down below to sense passing prey. Prey items include small fish fry and tadpoles, which are dragged into the bell to be eaten.
• **LIFE-CYCLE** After mating has taken place, the eggs are wrapped in silk and are then placed in the top of the diving bell.
• **OCCURRENCE** Europe and parts of Asia. In either slow-flowing or still water.

layer of air trapped by body hairs and flicked into bell by legs

"bell" of air held in place in vegetation by silk net

ARGYRONETA AQUATICA, the European Water Spider, even spends the winter in its bell-shaped tent. It adds extra silk to reinforce the structure and stays there until spring.

Length 0.7–1.5cm (⁹⁄₃₂–⅝in)	Feeding habits 🦗

Order ARANEAE	Family CTENIDAE	No. of species 600

WANDERING SPIDERS

These spiders are usually either grey or brown
in general coloration. The rear portion of their
carapace has a distinctive groove, running
lengthways. Most species are aggressive,
nocturnal hunters. They search for suitable
prey on the ground and then return to their
dark hiding places at dawn.
• LIFE-CYCLE Eggs are often laid in a silk
sac that the female carries under her body.
• OCCURRENCE Tropical and subtropical
regions. On the ground or on low-growing plants.
• REMARK The bites of some wandering
spiders can be dangerous to humans.

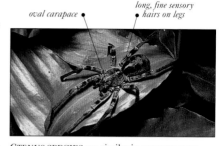

oval carapace •

long, fine sensory
• hairs on legs

CTENUS SPECIES are similar in appearance to
wolf spiders (see p.233). This drably coloured,
mottled specimen is from Africa.

Length 1.5–5cm (⅝–2in)		Feeding habits 🦗

Order ARANEAE	Family DIPLURIDAE	No. of species 250

FUNNEL-WEB SPIDERS

Mostly dark brown, these spiders have
six or eight eyes arranged in two groups
and a flat carapace. Their flat webs have
a funnel-shaped retreat that leads into
crevices in tree stumps, stones, and rocks.
• LIFE-CYCLE Females produce
tough, disc-shaped egg sacs that they
keep at the bottom of the retreat.
• OCCURRENCE Tropical and
subtropical regions of North America,
Africa, Asia, and Australia. In various
habitats, on the ground and in trees.

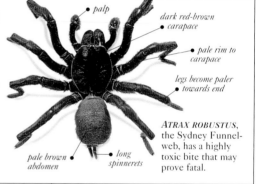

• palp

dark red-brown
• carapace

• pale rim to
carapace

legs become paler
• towards end

ATRAX ROBUSTUS,
the Sydney Funnel-
web, has a highly
toxic bite that may
prove fatal.

pale brown •
abdomen

• long
spinnerets

Length 0.6–2.8cm (¼–1⅛in)		Feeding habits 🦗

Order ARANEAE	Family DYSDERIDAE	No. of species 250

DYSDERID SPIDERS

Most of these spiders have six eyes, arranged roughly in a
circle. The chelicerae are often large, and the long fangs are
sharp enough to pierce tough cuticle. The abdomen may be
pinkish grey or patterned. Most are nocturnal and hunt on
the ground or make tubular silk nests in cavities in bark or
wood or among stones. In tube-nesting species, threads
radiating from the nest entrance trip up passing prey.
• LIFE-CYCLE Females may wrap their eggs in silk,
and the eggs are always kept inside a silk-lined retreat.
• OCCURRENCE Worldwide. In varied habitats, in natural
cracks and crevices in bark and wood and among stones.
• REMARK Tube-nesting species in the genus *Segestria*
are sometimes placed in a separate family.

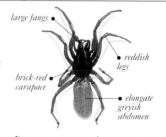

large fangs •

• reddish
legs

brick-red •
carapace

• elongate
greyish
abdomen

DYSDERA CROCATA is common
worldwide. Species in this genus
are known as woodlice-eating
spiders as these crustaceans form
the largest part of their diet.

Length 0.6–2.4cm (¼–1in)		Feeding habits

Order ARANEAE	Family ERESIDAE	No. of species 120

ERESID SPIDERS

These robust, quite hairy spiders have a large carapace and a square-fronted cephalothorax with eight eyes. Males may be brightly coloured. Some species make tubular webs in holes in the ground connected to funnel-shaped webs on the surface. Others make a web in shrubs.
• **LIFE-CYCLE** Females keep the egg sacs inside the retreat or carry them under their body.
• **OCCURRENCE** Parts of Africa, Europe, and Asia. In various habitats, in shrubs, and on the ground.

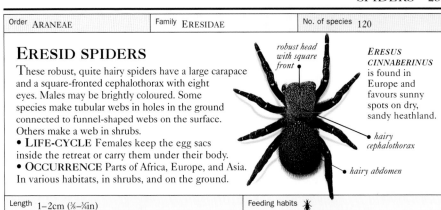

robust head with square front

ERESUS CINNABERINUS is found in Europe and favours sunny spots on dry, sandy heathland.

hairy cephalothorax

hairy abdomen

Length 1–2cm (⅜–¾in)	Feeding habits

Order ARANEAE	Family HETEROPODIDAE	No. of species 1,000

HUNTSMAN SPIDERS

Most species in this family are drably coloured, with mottled patterns. The carapace is typically as wide as it is long and, like the abdomen, is flattened in profile. The eight eyes are of equal size, four of them pointing forward from the front edge of the carapace. The legs can be very spiny and are often held out to the sides. Huntsman spiders hide under bark and stones or in vegetation during the day, or may be seen on tree trunks, and they hunt for prey at night. They are able to move sideways with great agility.
• **LIFE-CYCLE** Courtship can be quite complex. The female produces a silk egg sac that she hides under a stone or bark and guards until the eggs hatch, when she opens the sac to release the spiderlings.
• **OCCURRENCE** Worldwide, especially in tropical and subtropical regions. In various habitats, on the ground and on tree trunks.

long legs relative to body

eight eyes

drab coloration

olive-green legs with dark spines

broad, pale-rimmed carapace

male spider resting on tree bark

△ *HETEROPODA* SPECIES are often imported in crates of bananas. They have been known to bite warehouse workers but are not dangerous.

◁ *HETEROPODA VENATORIA* is a widespread species that is helpful to humans in tropical regions as it eats small scorpions and cockroaches.

Length 1–5cm (⅜–2in)	Feeding habits

Order ARANEAE	Family LINYPHIIDAE	No. of species 4,200

DWARF SPIDERS

As the common name implies, many of these spiders are small. The chelicerae are relatively large, with sharp teeth, and the legs have strong bristles. Males may have odd projections on the carapace, which may carry the eyes. Coloration varies from pale yellow to black and some have pale patches or banded legs. Many species attach non-sticky sheet webs to vegetation. Passing insects are knocked down on to the web, where the spider bites them from below with its chelicera and drags them under the sheet.

• **LIFE-CYCLE** Females may grip the male during mating. They attach egg sacs of various designs to plants, stones, and other surfaces.

• **OCCURRENCE** Worldwide, mostly in temperate areas. Among vegetation and stones in various habitats, such as woods, grassland, scrubland, and swamps. Dwarf spiders can travel vast distances by "ballooning" on silk threads.

• **REMARK** Another common name – "money spider" – comes from the myth that if a dwarf spider lands on you and is twirled around the head three times, good fortune will result.

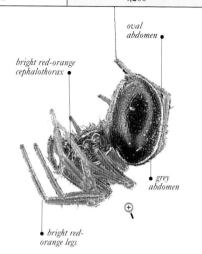

oval abdomen

bright red-orange cephalothorax

grey abdomen

bright red-orange legs

GONATIUM SPECIES are extremely common and several are widespread throughout the northern hemisphere. They are found in low vegetation or shrubs and prefer shady spots.

Length 0.1–1cm (¹⁄₃₂–³⁄₈in), most under 0.5cm (³⁄₁₆in)	Feeding habits

Order ARANEAE	Family LYCOSIDAE	No. of species 3,000

WOLF SPIDERS

These spiders vary from pale grey to dark brown with markings such as bands, stripes, white hairs, and black dots. The "head" area is often narrow, and the front two pairs of legs have many strong spines. Wolf spiders have four large eyes: the rear two face sideways and the two adjacent eyes face forwards. They also have four small eyes. These spiders have the excellent eyesight necessary for effective hunting, and most search for prey along the ground or among leaf-litter, usually at night.

• **LIFE-CYCLE** Courtship can be complex. Females of ground-active species carry egg sacs around with them, attached to their spinnerets. Burrowing species keep their egg sacs in a silk burrow. When the spiderlings hatch out, the mother may carry them around on her back.

• **OCCURRENCE** Worldwide, even in the Arctic. Widespread in varied habitats. Many are vital predators in fields, eating pests such as aphids, and some live in swamps, on plants, and on the surface of water.

distinctive black-and-white bands of hair

stout legs

egg sac attached to spinnerets

PARDOSA AMENTATA is common in Europe, where it prefers open habitats. This species can be quite variable in appearance – the abdomen may be either brown or grey, for example.

Length 0.4–4cm (⁵⁄₃₂–1½in)	Feeding habits

| Order ARANEAE | Family OONOPIDAE | No. of species 250 |

OONOPIDS

These spiders are often brightly coloured red, pink, orange, or pale yellow. The abdomen may have toughened plates on both dorsal and lateral surfaces. Most species have six eyes, grouped closely together, but some have only two or four eyes, or none at all. They do not spin webs but move about in leaf-litter after dark, preying on small insects. A few eat the remains of prey that they find in the webs of other spiders.
• LIFE-CYCLE The female produces very few eggs, enclosed in a silk sac inside her daytime retreat.
• OCCURRENCE Worldwide, mostly in forested, tropical regions. Some are found in houses.

six eyes, with a closely spaced central pair

abdomen paler than cephalothorax

OONOPS DOMESTICUS is found in European houses. In southerly areas, it can also be found out of doors. It moves in a highly distinctive way, interspersing short dashes with slow walking.

| Length 1–3mm (¹⁄₃₂–¹⁄₈in) | Feeding habits |

| Order ARANEAE | Family PHOLCIDAE | No. of species 350 |

DADDY-LONG-LEGS SPIDERS

The eyes of these spiders are arranged in two groups of three, with another pair in between. Typically, the carapace is very round. The legs are much longer than the body, giving a spindly appearance like that of harvestmen (see p.221). Coloration is grey, green, or brown, with pale legs. The tarsi are very long and flexible. These spiders make tangled, irregular webs and quickly wrap trapped prey in silk before biting it.
• LIFE-CYCLE Females produce about 15 to 20 eggs, which they wrap in a silk bundle and carry in their mandibles until they hatch.
• OCCURRENCE Worldwide. Many tropical species occur in caves or leaf-litter. In temperate regions, many species live in the dark corners of buildings.

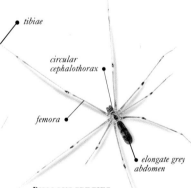

extremely long legs

tibiae

circular cephalothorax

femora

elongate grey abdomen

palps (in male)

dark "knees"

brown body

PHOLCUS SPECIES are common inhabitants of caves and rock crevices, especially in warmer regions.

PHOLCUS PHALANGIOIDES is now very common in buildings all over the world, but does not like cold temperatures.

| Length 0.3–1.4cm (¹⁄₈–⁵⁄₈in) | Feeding habits |

Order ARANEAE	Family PISAURIDAE	No. of species 550

NURSERY-WEB SPIDERS

These large, long-legged spiders are very similar to wolf spiders (see p.233), except that their eyes are smaller. Their body colouring varies from pale grey to dark brown, while the legs can be brown or white, and may have yellow bands. Rather than catch prey with webs, they run on the ground to hunt prey. The carapace is oval, with longitudinal markings. The common name refers to the protective web spun by the females for her young.

• **LIFE-CYCLE** The female carries her egg sac in her chelicerae. When the young are about to hatch out, many females spins a tent-like nursery web around the sac, among vegetation. She then guards her spiderlings.

• **OCCURRENCE** Worldwide. Widespread in various habitats on the ground and on the surface of still water or on aquatic plants.

abdomen with central and lateral stripes

broad, pale stripe down sides of carapace

brown legs with black spines

DOLOMEDES SPECIES are large, semi-aquatic spiders that catch tadpoles and fish fry as well as insects. Common across the northern hemisphere, they are able to jump on and off the water's surface.

Length 1–2.6cm (⅜–1in)	Feeding habits 🕷

Order ARANEAE	Family SALTICIDAE	No. of species 5,000

JUMPING SPIDERS

Most jumping spiders, so-called because they jump at prey, are drab in appearance, although tropical species can be brightly coloured with vivid markings. Four of the eight eyes form a row at the front of the carapace. The middle two are much larger than the rest, often resembling old-fashioned car headlights. Mostly daytime hunters, with excellent eyesight, they stalk prey to close range and then jump to seize them. A silk safety line ensures that they do not fall when stalking on vertical surfaces.

• **LIFE-CYCLE** Females usually lay eggs among vegetation, moss, bark, and stones, inside a large silk cell that they spin. They guard the eggs until they hatch.

• **OCCURRENCE** Worldwide, especially in warm regions. In a variety of habitats, including woods, grassland, heaths, and gardens. On walls, on the ground, in bushes, and often seen in sunny spots.

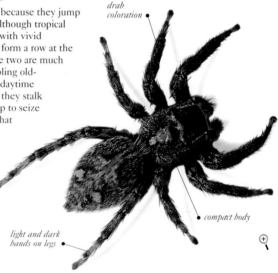

drab coloration

light and dark bands on legs

compact body

EUOPHRYS SPECIES are usually found under stones or near the ground on low-growing plants. Some specialize in hunting ants.

Length 0.2–1.6cm (¹⁄₁₆–⅝in)	Feeding habits 🕷

Order ARANEAE	Family SCYTODIDAE	No. of species 180

SPITTING SPIDERS

Typically cream- or yellow-brown with black markings, and with black-banded legs, the spitting spider has only six eyes, and the first pair of slender legs are usually longer than the others. At first glance, the carapace of the cephalothorax looks almost the same size as the abdomen. Seen in side view, the carapace is characteristically domed towards the rear, and the dome houses large glands that produce a sticky glue. This spider's common name comes from its unique prey-capturing technique. It does not spin webs, but uses a rapid, side-to-side movement of the chelicerae to "spit" two zig-zag streams of its glue at prey from close range, literally sticking it down.

• **LIFE-CYCLE** The female carries a pale and knobbly egg sac around underneath her body until the young emerge.

• **OCCURRENCE** Worldwide, except in Australia and New Zealand. Mostly in warm regions. Under rocks and in buildings.

• **REMARK** All the species in this family belong to the genus *Scytodes*.

pale brown legs with black bands •   *cephalothorax nearly same size as abdomen* •   *cream abdomen with dark, symmetrical bars and spots* •

SCYTODES THORACICA, native to North America and Europe, is a darkly marked spider that is often found inside buildings. The male is slightly smaller than the female.

Length 0.4–1.2cm (⁵⁄₃₂–¹⁄₂in)	Feeding habits 🐜

Order ARANEAE	Family SICARIIDAE	No. of species 100

SIX-EYED CRAB SPIDERS

Also known as brown spiders, because of their general body colour, most species have a violin-shaped mark on their carapace and a distinctive longitudinal groove. There are six eyes, arranged in three pairs. Both the body and legs have distinct hairs. These spiders make irregular, sticky, sheet-like webs.

• **LIFE-CYCLE** The females produce between 30 and 300 eggs per sac and keep the sacs out of the way, at the rear of the web. Some species live for several years, adding to their webs as they grow.

• **OCCURRENCE** Warm regions of North and South America, and also in Europe and Africa. In a wide variety of habitats, including woods, scrubland, citrus groves, gardens, and houses. In shady locations among rocks and bark, and sometimes in human dwellings.

• **REMARK** The bite of six-eyed crab spiders can be extremely dangerous, causing tissue degeneration.

long, slender legs •   *brown coloration* •   *quite broad carapace* •

LOXOSCELES RUFESCENS, a fiddle-back spider, may bite humans and produce unpleasant lesions that are slow to heal. It is common in Europe and has been introduced to Australia.

Length 0.6–1.8cm (¹⁄₄–³⁄₄in)	Feeding habits 🐜

Order ARANEAE	Family THERAPHOSIDAE	No. of species 800

TARANTULAS

Some larger tarantulas are also called bird-eating spiders.
These large, hairy spiders are usually pale brown to black, with
markings in shades of pink, red, brown, or black. The fangs bite
vertically, not horizontally. They have eight small eyes, grouped
together at the front of the carapace. Most tarantulas hunt on the
ground by night for arthropods and small vertebrates such as
frogs and mice. They use their large chelicerae to
crush their prey, pour digestive juices over the
body, and then suck up the resulting liquid.
• LIFE-CYCLE Some species live in trees,
whereas others make burrows in the ground. Females
lay a batch of eggs in the burrow. An egg sac can be
the size of a golf ball and contain 1,000 eggs.
The spiderlings stay in the burrow until
their first moult, after which they disperse to
find food and to make their own burrows.
• OCCURRENCE Worldwide,
especially in South America. In
subtropical and tropical areas,
in deserts, forests, and a variety
of open habitats.
• REMARK Many tarantulas live for
10–30 years and some are kept as pets.
Because many are so large, it is widely
assumed that their bites are fatal. Some
have potent venom but many do
not, relying on size to subdue
prey. The most poisonous
species are often
relatively small.

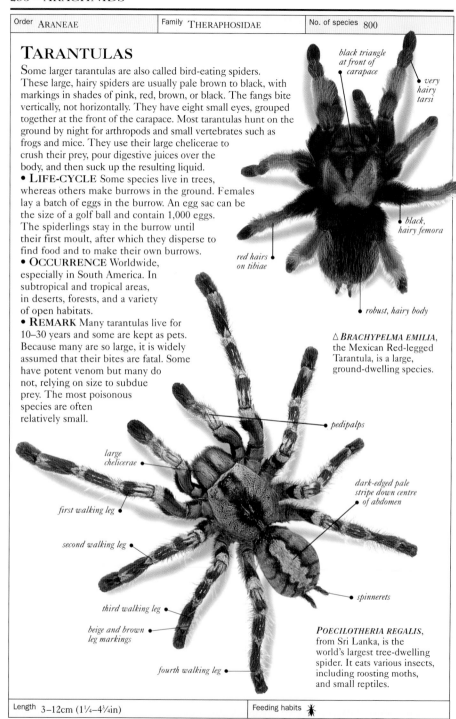

*black triangle
at front of
carapace*

*very
hairy
tarsi*

*black,
hairy femora*

*red hairs
on tibiae*

robust, hairy body

△ *BRACHYPELMA EMILIA*,
the Mexican Red-legged
Tarantula, is a large,
ground-dwelling species.

pedipalps

*large
chelicerae*

*dark-edged pale
stripe down centre
of abdomen*

first walking leg

second walking leg

third walking leg

*beige and brown
leg markings*

fourth walking leg

spinnerets

POECILOTHERIA REGALIS,
from Sri Lanka, is the
world's largest tree-dwelling
spider. It eats various insects,
including roosting moths,
and small reptiles.

Length 3–12cm (1¼–4¾in)	Feeding habits 🕷

| Order ARANEAE | Family THERIDIIDAE | No. of species 2,200 |

COMB-FOOTED SPIDERS

Also called cobweb spiders, these species are brown to black, often with markings and stout bristles on their hindlegs. The abdomen is very rounded. Most species are active at night, and some hunt for prey on the ground. They make irregular webs in foliage, cracks, crevices, and debris, or under buildings.
• LIFE-CYCLE Females produce about 200–250 eggs, attached to the web in a sac. After their first moult, the spiderlings make their own webs.
• OCCURRENCE Worldwide. In vegetation, under stones, in leaf-litter, and in and around buildings.
• REMARK The infamous widow spiders (including the notorious female American Black Widow Spider) and the Australian Red-back Spider belong to this family. These pea-sized black spiders have bright crimson markings on the underside of the abdomen. Their strong venom can kill, but a fast-acting anti-venom can be given by injection.

slender legs with a few fine spines

shiny all-black female

very rounded, globe-shaped abdomen

LATRODECTUS MACTANS, the venomous Black Widow Spider, is found in many tropical and subtropical countries.

| Length 0.2–1.5cm (¹⁄₁₆–⁵⁄₈in), most under 1cm (³⁄₈in) | Feeding habits |

| Order ARANEAE | Family THOMISIDAE | No. of species 2,500 |

CRAB SPIDERS

These spiders are named after their scuttling, sideways movements and squat shape, although some are elongate. The carapace is almost circular, and the short, often blunt-ended abdomen is frequently patterned. The first two pairs of legs, used to seize insect prey, are larger and spinier than the other two pairs. Many species are pink, yellow, or white to match the flowers on which they rest.
• LIFE-CYCLE Females keep their eggs in a flat sac, attached to plants, which they guard.
• OCCURRENCE Worldwide. Mainly in meadows and gardens, especially on flowers but also on plants and tree bark.

abdomen broad at rear

distinctive eye pattern

front two pairs of legs are the longest

rounded carapace

legs quite long

pale legs

△ ***MISUMENA VATIA*** is common throughout Europe and North America. The females are often found sitting on white or yellow flowers and can change from one colour to the other to blend in.

TIBELLUS OBLONGUS lies along grass blades with its head pointing downwards and seizes passing insects. It is widespread in damp meadows throughout Europe.

| Length 0.4–1.4cm (⁵⁄₃₂–⁵⁄₈in) | Feeding habits |

MYRIAPODS

PAUROPODS

T HERE ARE 5 families and 500 species in the order Pauropoda. These small, soft-bodied myriapods are typically pale in colour and have no eyes. The head carries a pair of branched antennae and weakly developed mouthparts. Located behind the head is a trunk, usually made up of between 9 and 11 segments. Hairs protrude from the upper surface of the head and trunk. The dorsal surfaces of the trunk segments (tergites) are fused together in pairs. Adult pauropods normally have 9 to 11 pairs of legs.

Pauropods live in soil or leaf-litter, where they are scavengers and eat fungal threads. A few species may be predacious. Reproduction begins with the male depositing a spermatophore that is picked up by the female.

Order PAUROPODA	Family PAUROPODIDAE	No. of species 450

PAUROPODS

These small myriapods have very pale, slender bodies and relatively long legs. The upper surface of the head and trunk carry long, pale, slender hairs, which have a sensory function. The antennae have four segments, and the last segment is divided into two branches. Despite their small size, pauropods can move quite quickly through soil crevices.
• LIFE-CYCLE In most species, the females deposit their eggs, either singly or in small batches, in the soil or among rotting plant matter. The complete life-cycle, from egg to adulthood, takes about three or four months.
• OCCURRENCE Worldwide. In various habitats, especially woodland, in soil, leaf-litter, and rotting logs, in damp, draught-free places. Some species are very widespread.
• REMARK This family contains most of the known species of pauropod.

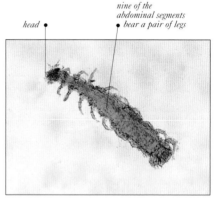

head •

nine of the abdominal segments • *bear a pair of legs*

ALLOPAUROPUS DANICUS is found throughout the northern hemisphere. It has long, fine body hairs that are not visible on this slide-mounted specimen.

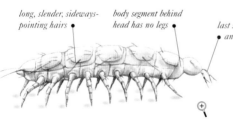

long, slender, sideways-pointing hairs •

body segment behind head has no legs •

last segment of • *antennae branched*

PAUROPUS SPECIES may resemble very small, grub-like centipedes (see pp.240–41) in general appearance. They can be very abundant in forest humus and leaf-litter.

Length 0.5–2mm (¹⁄₆₄–¹⁄₁₆in)		Feeding habits

SYMPHYLANS

A RELATIVELY SMALL GROUP, the order Symphyla contains just 2 families and 175 species. These small, soft-bodied myriapods are usually white or pale brown or grey. The head bears a pair of long and thread-like antennae and three pairs of mouthparts. There are no eyes. Behind the head is a trunk, made up of 14 segments. In adults, the first 12 segments typically have a pair of six-segmented legs. The last trunk segment has a pair of short spinnerets, similar in appearance to the cerci found in insects, but whose function is to produce fine silk. Unlike pauropods (see p.238), symphylans have trunk segments with tergites that are not fused together in pairs. Some of the tergites are also doubled in most species. This higher number of unfused tergites gives symphylans a great deal of flexibility.

Symphylans live in soil or in leaf-litter, often in vast numbers, where they feed on plant material. Fertilization is indirect, with the female of the species picking up a spermatophore that has been deposited by the male. It is not clear whether symphylans are more closely related to centipedes (see pp.240–41) or to pauropods (see p.238) and millipedes (see pp.242–43); they have characteristics that are suggestive of both groups.

Order SYMPHYLA	Family SCUTIGERELLIDAE	No. of species 100

SCUTIGERELLIDS

These are short and stout symphylans with tough tergites. They are typically pale grey, straw-coloured, or white. Scutigerellids are highly flexible and are also able to run very rapidly, twisting and turning their way through tiny crevices to escape predators.

• LIFE-CYCLE The females use their mouths to pick up the stalked spermatophore deposited on the ground by the males, keeping the sperm in special pouches in their mouth. The females also use their mouthparts to remove eggs from their genital opening and then smear the eggs with sperm and stick them to a plant or position them in a crevice in the soil. A cluster of about 30 eggs may be laid at one time. When they hatch out, the first-stage young have six pairs of legs. They gain additional body segments and legs with each successive moult. Some species of symphylan can live for three or four years.
• OCCURRENCE Worldwide. In varied habitats, in soil and leaf-litter.
• REMARK Many species are root-feeders and can cause losses in seedling and tuber crops.

relatively long antennae

flexible body

white coloration

SCUTIGERELLA IMMACULATA is commonly found in gardens and so is popularly known as the Garden Symphylan. Like several other symphylan species, it may become a minor pest in greenhouses.

Length 3–8mm (⅛–⁵⁄₁₆in)	Feeding habits

CENTIPEDES

THE 4 ORDERS, 22 FAMILIES, and 3,000 species in the class Chilopoda – the centipedes – are predacious and hunt mostly at night. They use poison claws to kill prey, which in some cases is as large as mice. Centipedes are long and usually flat, with a head that bears mouthparts and segmented antennae. The trunk has at least 16 segments, most of which carry a pair of legs, the last being the longest. Usually yellow or brown, the body may be green- or red-tinged and is covered with fine sensory hairs.

Courtship is common. Males drop sperm on the ground that the female picks up. Eggs are laid singly or brooded underground in batches. Common in a range of habitats, many centipedes are found in temperate regions but most are native to subtropical and tropical areas.

Order GEOPHILIDA	Family GEOPHILIDAE	No. of species 200

GEOPHILIDS

The body of these straw-coloured to brown centipedes is long, slender, and made up of at least 35 segments. The legs are short.
• **LIFE-CYCLE** Females typically lay eggs in soil, but some lay eggs under bark and use their poison claws to position and turn them. The young have most of their legs before they hatch.
• **OCCURRENCE** Worldwide. In most habitats, forming burrows in soil, leaf-litter, and debris.
• **REMARK** The name means "earth-loving".

thread-like body
short legs

***GEOPHILUS* SPECIES** females brood egg masses underground, coiling their bodies around them. There are up to 181 pairs of legs, depending on the species.

Length 1–5cm (³⁄₈–2in)	Feeding habits 🐜

Order LITHOBIIDA	Family LITHOBIIDAE	No. of species 1,500

LITHOBIIDS

Most lithobiids are red-brown, but some are brightly coloured. The tough, flat body has 15 pairs of legs, the last two pairs being the longest. Some species have eyes consisting of up to 34 ocelli.
• **LIFE-CYCLE** Females lay round eggs singly, one or two days apart and push them into the soil. Newly hatched young have seven pairs of legs and gain a leg-bearing trunk segment at each moult.
• **OCCURRENCE** Worldwide, especially in temperate parts of the northern hemisphere. In most habitats, in cracks and crevices.

poison claws • broad head •
groups of ocelli on each side of head •
dark brown trunk segments, often with purple tinge •

antennae one third body length •
chestnut-brown coloration •
last pair of legs longer than other pairs •

▷ ***LITHOBIUS VARIEGATUS*** is common among the litter of deciduous woods. It climbs trees in search of food.

◁ ***LITHOBIUS FORFICATUS*** is found under stones and bark throughout the northern hemisphere.

• dark central line
• alternate light and dark bands on rear legs

Length 0.6–3.8cm (¼–1½in)	Feeding habits 🐜

Order SCOLOPENDRIDA	Family SCOLOPENDRIDAE	No. of species 400

SCOLOPENDRIDS

These robust centipedes are typically brightly coloured and may be yellow, red, orange, or green, often with dark stripes or bands. They have 21 or 23 pairs of legs and usually four ocelli on either side of the head.
• LIFE-CYCLE Females may burrow under soil, leaf-litter, rocks, or tree bark to lay their eggs. The young have all their legs before they hatch.
• OCCURRENCE Worldwide, especially in subtropical and tropical regions. In soil, leaf-litter, or cracks and crevices. Some climb shrubs and trees.
• REMARK At 30cm (12in) long, *Scolopendra gigantea*, from South America, is the world's largest centipede. Bites from members of this genus, especially smaller species, can cause great pain, fever, and vomiting.

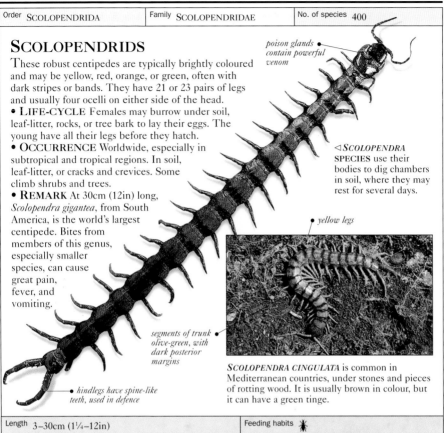

poison glands contain powerful venom

◁ *SCOLOPENDRA* SPECIES use their bodies to dig chambers in soil, where they may rest for several days.

yellow legs

segments of trunk olive-green, with dark posterior margins

hindlegs have spine-like teeth, used in defence

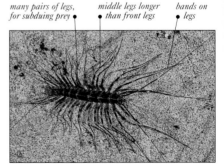

SCOLOPENDRA CINGULATA is common in Mediterranean countries, under stones and pieces of rotting wood. It is usually brown in colour, but it can have a green tinge.

Length 3–30cm (1¼–12in)	Feeding habits 🐜

Order SCUTIGERIDA	Family SCUTIGERIDAE	No. of species 150

SCUTIGERIDS

These fairly short centipedes are brown with paler markings. They have 15 pairs of very long legs. Those at the rear are much longer than those at the front. The body is kept straight by rigid, overlapping "plates" (tergites). The round head bears large compound eyes.
• LIFE-CYCLE Females lay one egg at a time – on the ground, in crevices, and among debris – and show no brooding behaviour. Newly hatched young have seven pairs of legs and gain a leg-bearing trunk segment at each moult.
• OCCURRENCE Worldwide, but mainly in warm regions. In various habitats, including buildings and caves.
• REMARK *Scutigera coleoptrata*, the House Centipede, can run at 40cm (16in) per second.

many pairs of legs, for subduing prey • *middle legs longer than front legs* • *bands on legs*

SCUTIGERA SPECIES are fast-running insectivores. A few species are commonly found inside buildings all year round and also in the open during summer.

Length 1–5cm (⅜–2in)	Feeding habits 🐜

MILLIPEDES

WITHIN THE CLASS DIPLOPODA, there are 13 orders, 115 families, and 10,000 species. The majority of millipedes are dull in colour and are slow-moving, with tough, cylindrical bodies, strong mandibles, and seven-segmented antennae. The first four trunk segments have no legs; the other segments bear two pairs of legs. Although the common name implies that they have 1,000 legs, most have far fewer and none has more than 750. Males twist their bodies around the females to transfer sperm. A female lays her eggs inside soil nests, and most young hatch out with six legs, gaining legs and body segments as they moult.

Most millipedes live in soil, leaf-litter, or debris, eating mainly rotting organic matter or fungi. To protect themselves, they roll up or produce toxic chemicals.

Order GLOMERIDA	Family GLOMERIDAE	No. of species 200

PILL MILLIPEDES

The trunk of this millipede is made up of 13 segments. The shape of the dorsal plates covering each of these segments allows it to roll up into a tight ball with its head tucked in. The small species are drably coloured, but large species can be brightly marked. Adults have 15 pairs of legs.
• LIFE-CYCLE As with all millipedes, the females typically lay their eggs in nests that they have made in the soil.
• OCCURRENCE Widespread throughout both warm and cool temperate regions of the northern hemisphere. In soil and caves.

distinctive body segment just behind head

broad body

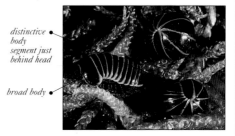

GLOMERIS MARGINATA can easily be confused with pill woodlice (see p.212) when completely rolled up, although it has a shinier body.

Length 0.2–2cm (¹⁄₁₆–¾in)		Feeding habits

Order JULIDA	Family JULIDAE	No. of species 450

CYLINDER MILLIPEDES

As their common name implies, these millipedes have very rounded bodies. They are usually dull in colour, although a few species may have either red or pale cream or brown spots.
• LIFE-CYCLE Like all millipedes, the females of this family typically lay their eggs in nests in soil. There are usually seven nymphal stages.
• OCCURRENCE Mainly in the northern hemisphere, especially Europe and Asia. In a variety of habitats, in soil and leaf-litter and underneath stones and rotting wood. Some species may be found in caves and at high altitudes.

stiff, rounded body segments

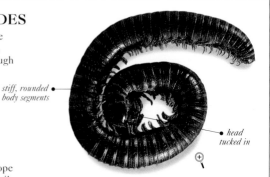

head tucked in

***JULUS* SPECIES** are typical of millipedes in using their strong, stiff, body segments and many legs to push themselves through soil and leaf-litter.

Length 0.8–8cm (⁵⁄₁₆–3¼in)		Feeding habits

Order POLYDESMIDA	Family POLYDESMIDAE	No. of species 200

FLAT-BACKED MILLIPEDES

Also known as plated millipedes, members of this
family are less rounded than other millipedes.
They have no eyes and are generally dull-coloured,
with expansions on the top part of their body
segments that project horizontally. In most
species, many of the body segments
contain glands that can produce toxic
chemicals to deter predators.
• **LIFE-CYCLE** Like other
millipedes, the females lay eggs in
nests made in soil. Young stages
live in the soil, whereas older
stages and adults forage for
food on the surface.
• **OCCURRENCE** Northern
hemisphere. In woodland
leaf-litter. Some species are
found in caves.
• **REMARK** Some Californian
species in the genus *Motyxia* are
luminescent. There are other
species of flat-backed millipede
that can live under water.

antennae

legs

shiny, pitted surface

paired legs

backward-pointing lateral expansions

brown coloration

white or yellow-orange margins to expansions in some species

lateral expansion to body segment

POLYDESMUS **SPECIES** may be
mistaken for centipedes because
of their flat shape. However,
members of this genus can easily be
distinguished by their paired legs –
obvious in the right-hand specimen.

Length 0.5–3.2cm (³⁄₁₆–1¼in)	Feeding habits

Order SPIROSTREPTIDA	Family SPIROSTREPTIDAE	No. of species 800

SPIROSTREPTIDS

Some of these millipedes are
brightly coloured, although
most are dull. The body
segments are typically smooth,
but they may be pitted.
• **LIFE-CYCLE** Eggs are
laid in soil nests.
• **OCCURRENCE** Mostly
southern hemisphere, in
subtropical and tropical regions.
In forests but also in semi-arid
areas. Some are found in trees.
• **REMARK** This family
contains the world's largest
millipede, the African species
Graphidostreptus gigas.

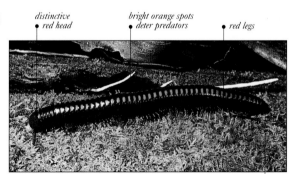

distinctive red head

bright orange spots deter predators

red legs

red tail

SPIROSTREPTUS **SPECIES** are giant, brightly
coloured tropical millipedes. The bright
coloration warns predators that they can secrete
toxic chemicals from their body segments.

Length 1.2–28cm (½–11in)	Feeding habits

GLOSSARY

Many of the terms described here are illustrated in the introduction (see pp.6–45). Words in **bold** type are defined elsewhere in the glossary.

• **ABDOMEN**
The rearmost of the three main segments of a typical **insect**. The head and **thorax** are the other two main segments.

• **AMETABOLOUS**
Developing without obvious **metamorphosis**.

• **ANTENNA (*pl.* ANTENNAE)**
One of a pair of mobile appendages on the heads of insects and certain other invertebrates; they respond to taste and touch.

• **ARACHNID**
An **arthropod** with a body that is divided into two main parts (the **cephalothorax** and **abdomen**), and that has **chelicerae**, four pairs of **walking legs**, and **simple eyes**.

• **ARTHROPOD**
A member of the phylum Arthropoda. Arthropods have segmented bodies with jointed limbs and a tough **exoskeleton**.

• **CARAPACE**
The toughened protective **dorsal** plate covering the **cephalothorax** of some **arthropods**.

• **CASTE**
A physically or behaviourally specialized group within an insect colony.

• **CEPHALOTHORAX**
The body section in **arachnids** and **crustaceans** made up of the fused head and **thorax**.

• **CERCUS (*pl.* CERCI)**
One of a pair of "tails" extending from the end of the **abdomen** in some **insects**, often with a sensory function.

• **CHELICERA (*pl.* CHELICERAE)**
The first of six pairs of appendages on the **cephalothorax** in **arachnids**. Chelicerae are pincer- or fang-like and used mainly for handling prey.

• **CHRYSALIS**
The **pupa** of a butterfly.

• **COCOON**
A protective case made by the fully grown **larva** of many **insects** just before **pupation**. It is composed partly or completely of silk.

• **COMPLETE METAMORPHOSIS**
See **Metamorphosis**.

• **COMPOUND EYE**
The large eye, made of numerous separate facets (called ommatidia) found in many insects.

• **CREMASTER**
The hooked appendage on the rear end of a **chrysalis**.

• **CRUSTACEAN**
An **arthropod** with jaws and gills. Crustaceans are typically marine; the main terrestrial examples are species of woodlice (see p.212).

• **CUCKOO**
An insect that uses the food stored by another to rear its own young.

• **CUTICLE**
See **Exoskeleton**.

• **DORSAL**
Relating to the upper surface, or "back", of a structure or organism. *See also* **Ventral**.

• **DORSO-VENTRALLY FLATTENED**
Flattened from top to bottom (rather than side to side).

• **DRONE**
A male honeybee, whose sole function is to mate with the queen.

• **ECOSYSTEM**
A web or linked network of relationships and interactions between living things and their environment.

• **ECTOPARASITE**
A **parasite** that lives on the outside of its **host**, feeding on it without killing it. Notable examples include lice (see p.83) and fleas (see p.135).

• **ECTOPARASITOID**
A **parasitoid** that lives on the outside of a **host**, feeding on it and killing it in the process.

• **ELYTRON (*pl.* ELYTRA)**
The rigid forewing of a beetle, which protects the hindwing.

• **ENDOPARASITE**
A **parasite** that lives on the inside of a **host**, feeding on the host but not necessarily killing it.

• **ENDOPARASITOID**
A **parasitoid** that lives on the inside of a **host**, feeding on it and killing it in the process.

• **EXOSKELETON**
The protective or supporting structure (cuticle) covering the body of an arthropod.

• **EYESPOT**
An eye-like marking, as on the wings of certain butterflies.

• **FEMUR (*pl.* FEMORA)**
The third segment of the leg (away from the body), situated just above the **tibia**. The femur is often the largest segment of the leg.

• **FURCULA**
The forked, **abdominal** jumping organ of springtails (see p.207).

• **GALL**
An abnormal outgrowth on various parts of a plant, caused by an insect or other organism (the gall-former). Aphids (see p.99) and gall wasps (see p.196) are some of the major gall-forming insects.

• **GILL**
The respiratory organ in many aquatic animals, including some insect **nymphs**.

• **GRUB**
The short, legless larva of certain insects, especially beetles.

• **HALTERE**
One of a pair of small, club-shaped organs that help two-winged flies (see p.136) to maintain balance while flying. Halteres have evolved from what were once hindwings.

• **HAPLODIPLOIDY (*adj.* HAPLOID)**
A fertilization process in some insects, in which fertilized eggs produce females and unfertilized eggs produce males.

• **HEMIMETABOLOUS**
Having **incomplete metamorphosis**.

• **HEXAPOD**
An **arthropod** with six legs.

• **HOLOMETABOLOUS**
Having **complete metamorphosis**.

• **HONEYDEW**
The carbohydrate-rich liquid excrement of sap-feeding species such as aphids (see p.99).

• **HOST**
An organism that is attacked by a **parasite** or **parasitoid**.

• **HYPERPARASITOID**
A **parasitoid** that uses another parasitoid as a **host**.

• **INCOMPLETE METAMORPHOSIS**
See **Metamorphosis**.

• **INSECT**
An **arthropod** and **hexapod**, typically with a segmented body that is divided into three segments. Most insects also have antennae and one or two pairs of wings.

• **INSTAR**
The stage in an insect's life-cycle between any two moults. The adult stage is the final instar.

• **LARVA** (*pl.* **LARVAE**)
The immature stage of an **insect** that undergoes **complete metamorphosis**.

• **LEAF-LITTER**
The layer of fallen leaves that is home to many **arthropods**.

• **LEAF-MINER**
A **larva** that burrows inside leaves, often leaving distinctively shaped tunnels, known as mines.

• **MANDIBLES**
The jaws of an insect. They may be toothed and used for biting, or they may be modified for piercing, as in mosquitoes (see p.138).

• **METAMORPHOSIS**
The transformation in a series of stages from an immature insect into an adult. In many insects, these stages form a **complete metamorphosis**, where the young look very different to the adults – as in beetles (see p.109) or moths and butterflies (see p.158). In complete metamorphosis, the immature stages are called **larvae**. The scientific name for complete metamorphosis is holometaboly. In other insects, there is an **incomplete metamorphosis**, where the young look like smaller versions of the adults – for example in mayflies (see p.48) and bugs (see p.85). The young of insects that develop by incomplete metamorphosis are called **nymphs**. The scientific name for incomplete metamorphosis is hemimetaboly.

• **MOULT**
To shed the outer covering of the body (the **exoskeleton**).

• **NAIAD**
The aquatic **nymph** of certain insects, notably dragonflies (see p.51).

• **NYMPH**
The immature stages of those **insects** that develop by incomplete or gradual **metamorphosis**.

• **OCELLUS** (*pl.* **OCELLI**)
A simple, light-receptive organ on the head of many **insects**. Three ocelli are often arranged in a triangular formation on the top of the head. Also called a simple eye.

• **OVIPOSITOR**
The egg-laying tube of many female **insects**. It may be hidden or highly conspicuous.

• **PALPS**
A pair of finger-like sensory organs that arise from the mouthparts of **arthropods**.

• **PARASITE** (*adj.* PARASITIC)
A species that lives off the body or tissues of another species – the **host** – without causing the host's death. *See also* **Ectoparasite** and **Endoparasite**.

• **PARASITOID**
A species that lives off the body or tissues of another species – the **host** – and causes the host's death. *See also* **Ectoparasitoid, Endoparasitoid,** and **Hyperparasitoid**.

• **PARTHENOGENESIS** (*adj.* PARTHENOGENETIC)
Reproduction without fertilization.

• **PEDIPALPS**
The second of six pairs of appendages on the **cephalothorax** of some **arachnids**. They may be used for walking or transferring sperm, but in some groups they are large and used for killing and handling prey.

• **PHEROMONE**
A chemical produced by animals in order to affect the behaviour of other animals – for example to attract a mate or deter predators.

• **PREDATOR** (*adj.* PREDACIOUS)
An animal that eats other animals.

• **PROBOSCIS**
The elongate mouthparts of certain insects, adapted for sucking food.

• **PROLEG**
An unsegmented leg on a **larval insect** – for example, one of the short legs on a caterpillar's **abdomen**.

• **PRONOTUM**
The **dorsal** covering over the first segment of the **thorax**.

• **PROTHORAX**
The first of three segments forming an insect's thorax. The other two segments are the mesothorax and the metathorax.

• **PTEROSTIGMA**
A toughened, often darkened, area on the front margins of the wings of many **insects**, notably dragonflies (see p.51). Also called a stigma.

• **PUPA**
The stage during which tissues are rearranged to form an adult body in **insects** that develop by complete **metamorphosis**. A pupa does not feed and is usually immobile.

• **PUPATE**
To turn into a **pupa**.

• **ROSTRUM**
The slender, sucking mouthparts of bugs (see p.85) or the elongate part of the head of weevils (see p.117) or scorpionflies (see p.133).

• **SIMPLE EYE**
See **Ocellus**.

• **SOLITARY**
Not occurring in gregarious or social groups.

• **SPERMATOPHORE**
A structure or "packet" produced by some **arthropods** to contain and transfer sperm to the female.

• **SPINNERET**
A moveable, conical structure at the end of a spider's **abdomen**, through which silk is extruded. There are typically three pairs of spinnerets.

• **SPIRACLE**
The breathing holes of insects, leading to the internal respiratory system.

• **STERNUM**
The **ventral** surface of an **arthropod** body segment – for example, the "breast-plate" of a scorpion.

• **STIGMA**
See **Pterostigma**.

• **STING**
The modified **ovipositor** of some insects in the order Hymenoptera (see p.178), used for defence.

• **TARSUS** (*pl.* **TARSI**)
The "foot" (or last leg segment) of an **insect**, which is made up of a variable number of segments called tarsomeres.

• **TELSON**
The "tail" or final segment of the **abdomen** of some **arachnids** and **crustaceans**.

• **THORAX**
The middle segment of the three segments that make up an insect's body (the other two being the head and the **abdomen**). The wings and legs are attached to the thorax.

• **TIBIA** (*pl.* **TIBIAE**)
The leg segment that is located between the **femur** and the **tarsus**.

• **VENTRAL**
Relating to the underside or lower surface of a structure or organism.

• **VESTIGIAL**
Having attained a simple structure and reduced size and function during the evolution of the species.

• **WALKING LEGS**
Legs used for walking as opposed to other purposes, such as killing and handling prey or transferring sperm to a mate.

INDEX

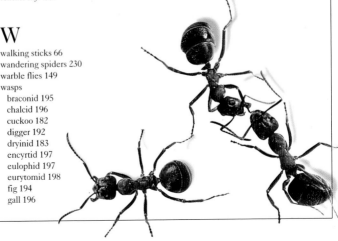

ACKNOWLEDGMENTS

THE AUTHOR would especially like to thank Darren Mann for sharing his extensive knowledge of insects. Other colleagues at the Oxford University of Natural History answered many questions and helped in various ways, notably Dr John Ismay, Dr Adrian Pont, Mr Christopher O'Toole, Professor Steve Simpson, Dr Derek Siveter, Professor David Spencer-Smith, Dr Matthew Wills, Dr Kwang-sun Cho, and Dorothy Newman. I am also grateful to Dr John Noyes, Dr Malcolm Scoble, and Dr Zhi-Quang Zhang at the Natural History Museum in London, Peter Smithers of the University of Plymouth, Dr John Deeming of the National Museums and Galleries of Wales, Dr Eugene Marais and Dr Eryn Griffin of the National Museum of Namibia, and Dr Frank Rodovsky and Dr Barry O'Connor of Oregon State University and University of Michigan respectively. Special thanks to Lois.

DORLING KINDERSLEY would like to thank Richard Hammond and Sean O'Connor for their invaluable editorial assistance. Thanks also to Peter Cross, Steve Knowlden, and Elaine Hewson for their design expertise.

STUDIO CACTUS would like to thank Sharon Moore, Ann Thompson, and Amelia Freeman for design assistance. Polly Boyd, Nicola Hodgson, Irene Lyford, Amanda Hess, Christine Davis, and Jane Baldock for editorial assistance. Thanks to Douglas Brown for compiling the index. Thanks also to Melanie Brown.